Lecture Notes in Mathematics

A collection of informal reports and seminars
Edited by A. Dold, Heidelberg and B. Eckmann, Zürich

125

Symposium on Automatic Demonstration

Held at Versailles/France, December 1968

Edited by
M. Laudet, IRIA, Rocquencourt/France,
D. Lacombe, L. Nolin and M. Schützenberger, Faculté des Sciences, Paris/France

Springer-Verlag
Berlin · Heidelberg · New York 1970

Lecture Notes in Mathematics, Vol. 125

ERRATA

Contribution:

REFINEMENT THEOREMS IN RESOLUTION THEORY by David Luckham

p. 170 In the definition of $\tilde{R}_2$, λ and τ are simplest (or most general) substitutions such that $A\lambda$ is a merge or $B\tau$ is a merge.

p. 180 Definition (ii) should read:

(ii) $R(\mathfrak{U}|K)$ = df. the subset of $R(\mathfrak{U})$ consisting of those clauses having an instance which contains only terms in $K(S)$.

p. 183 line 8 from the bottom: "the clause..." should be "a clause..."

p. 185 line 2: Tr(a) should be Tr(A).

Ce livre contient la plupart des exposés qui ont été présentés lors du Colloque International sur la Démonstration Automatique, organisé par l'Institut de Recherche d'Informatique et d'Automatique, en Décembre 1968, à Rocquencourt, France.

This book contains the greater part of the Confcrences which have been given during the international symposium on Automatic Demonstration organised in december 1968 by the French Institut de Recherche d'Informatique et d'Automatique, at Rocquencourt France.

Ces textes ont été dactylographiés par Mademoiselle HERNANDEZ, du C.N.R.S. à Paris.

These texts have been typed by Miss HERNANDEZ, C.N.R.S. in Paris.

SYMPOSIUM ON AUTOMATIC DEMONSTRATION

COLLOQUE DEMONSTRATION AUTOMATIQUE

CONTENTS

List of Contributors

Laudet, Michel: Domaine de Voluceau, I.R.I.A., Rocquencourt/France

Arnold, André: Faculté des Sciences, Lille/France

de Bruijn, N.G.: Technological University, Eindhoven / Netherlands

Engeler, Erwin: Froschungsinstitut für Mathematik, ETH, Zürich / Swiss and University of Minnesota, Minneapolis,MN/USA

Fraissé, Roland: Faculté des Sciences, Marseille / France

Grzegorczyk, Andrzej: Polish Academy of Sciences, Mathematical Institute, Warszawa / Poland

Hao Wang: Rockefeller University, New York City, NY / USA

Kowalski, Robert: Mathematics Unit, University of Edinburgh, Edinburgh / Scotland

Kreisel,G. : Stanford University, Dept. of Mathematics, Stanford, CA / USA and Université de Paris, Faculté des Sciences, Paris / France

Loveland, D.W.: Carnegie-Mellon University, Pittsburgh, PA / USA

Luckham, David: Computer Science Department, Stanford University, Standford, CA / USA

Pawlak, Z.: Institute of Mathematics, Warszawa University, Warszawa / Poland

Pitrat, Jacques: Institut Blaise Pascal (C.N.R.S.), Paris / France

Prawitz, Dag: Lunds Universitet, Lund/Sweden

Robinson, G. : Stanford Linear Accelerator Center, Stanford, CA / USA

Scott, Dana : Dept. of Mathematics,Stanford University,Stanford,CA/USA

Wos, L. : Argonne National Laboratory, Argonne, IL / USA

ALLOCUTION d'OUVERTURE M. LAUDET

Il serait présomptueux pour quiconque de traiter devant vous des problèmes hautement techniques autour desquels se regroupent vos travaux, et le généraliste que je suis s'estime mal qualifié pour être autre chose qu'un auditeur passionné à cette réunion.

Cependant, puisqu'il est d'usage qu'une allocution d'ouverture prépare les voies aux exposés plus ardus et puisque cette tâche me revient, je me tournerai vers les nombreux participants qu'a attiré votre renom pour leur rappeler comment le sujet de ce Colloque se place dans l'axe même du développement de la mathématique. Ceci naturellement je le ferai en fonction de ce qui est l'un des buts de notre Institut et de mes intérêts propres, à savoir la mathématique en tant que science des calculs.

Distinguons, si vous le voulez bien, trois niveaux:

- la construction;
- la décision;
- la semi-décision;

que l'on pourrait aussi bien associer schématiquement à des étapes historiques ou à des prises de consciences de rigueur et de possibilité.

1) <u>Les Constructions</u>: C'est essentiellement le niveau prémathématique des sciences égyptienne et babylonienne dont le papyrus de Rhind et les tablettes gravées nous apportent le témoignage.

Ici, il s'agit surtout comme on l'a tant répété d'un recueil de recettes - souvent ingénieuses, rarement systématiquesdont le but déclaré est de résoudre les problèmes numériques que requiert la technologie de l'époque: calculs cadastraux, calculs astronomiques, problèmes de partage. Le trait le plus frappant pour un moderne est l'absence de discussion des limites de validité. Le centre de celles-ci étant indiqué au moyen d'exemples typiques.

Cette approche réapparaîtra souvent dans l'histoire aux frontières des domaines déjà organisés en science rigoureuse et il faut lui rattacher les premières ébauches de la théorie des nombres de Diophante, de l'algèbre de Tartaglia et Jérôme Cardan, les "problèmes plaisants et délectables" de Bluchet de Meziriac.

Plus généralement nous verrions volontiers cette même méthode empirique, voire expérimentale, dans nombre de travaux directement inspirés des problèmes pressants de la physique et nous y saluons la source sans cesse renouvelée d'une inspiration et d'un guide pour les recherches des mathématiciens proprement dits.

Nous sommes sûrs, d'ailleurs, que les moyens puissants de l'Informatique y ont un grand rôle à jouer même dans des branches fort éloignées des applications traditionnelles.

2) La décision: Au second niveau se situe le calcul dans des théories décidables ou que l'on considère comme telles. C'est là naturellement que s'effectue la plus grande partie de l'activité des mathématiciens appliqués:

Par exemple, bien que les structures sous jacentes soient infinies la méthode de Jacobi pour diagonaliser une matrice conduit à un processus décidable chaque fois qu'a été fixée à l'avance la précision, au demeurant arbitraire, dont l'obtention provoquera l'arrêt des calculs. Il en serait de même de la recherche numérique des racines intégro différentielles. Dans d'autres cas, le problème est décidable parce que les structures en cause sont finies; comme c'est le cas des problèmes impliquant les algèbres de Boole ou le calcul des propositions.

Dans d'autres cas enfin ce sont des théories difficiles qui nous ont révélé le caractère décidable de toute une classe de problème. Par exemple, la théorie de Tarski a montré qu'il en était ainsi de l'algèbre et de la géométrie élémentaire et que l'on pouvait répondre de façon algorithmique à la question: un système donné d'équations et d'inéquations admet-il ou non une solution réelle ?

Vous savez mieux que moi l'importance et l'état actuel de ces recherches. Vous savez aussi combien certains de ces algorithmes sont longs et complexes dès qu'on s'écarte tant soit peu des cas les plus élémentaires. Permettez-moi une fois encore d'insister sur l'influence qu'ont et qu'auront les ordinateurs tant sur le plan pratique que sur le plan théorique.

Aurait-on songé à élaborer la théorie de la programmation linéaire en nombre entier si n'avaient existé les moyens matériels de résoudre effectivement des classes de problèmes dans une gamme entièrement inaccessible au calcul manuel?

Aurait-on songé à étudier comparativement l'efficacité des algorithmes si leur domaine d'application s'était limité à la marge étroite des cas à peu près triviaux qui peuvent être abordés sans machine?

3) Les théories indécidables: Toutefois nous le savons bien les théories les plus intéressantes sont indécidables. Le théorème de Gödel apporte à l'humaniste le message le plus réconfortant: aucune machine ne peut remplacer l'esprit humain. Rien ne nous retient donc de chercher à cerner de plus près la partie irréductible de son activité créatrice et pour cela deux voies divergentes se présentent à nous:

- Tenter de reproduire le plus fidèlement possible les démarches les plus élémentaires de l'esprit afin de pouvoir les extrapoler;
- Formuler et expérimenter des méthodes de semi-décision. Si la proposition dont nous voulons calculer la valeur de vérité est vraie, nous obtiendrons la réponse en un nombre fini d'étapes. Sinon, l'algorithme fonctionnera indéfiniment car, dans tous les cas, ce qu'il cherche est un contre exemple.

Telle est la méthode imaginée par Herbrand dès 1930 et dont il a fallu près de 30 ans avant de commencer même à en explorer les conséquences.

Certes, dès l'abord la longueur et l'opacité des calculs devaient décourager toute tentative d'expérimentation à la main et une fois encore ce sont les ordinateurs qui ont stimulé les recherches.

Ces deux voies d'ailleurs ont été explorées simultanément et les premiers résultats obtenus ont été, je crois, communiqués à un large public pour la première fois à Paris, en 1959, au premier Congrès de l'IFIP.

Dans la voie de la simulation, Gelernter, certains de vous s'en souviennent peut être, s'était efforcé de retracer artificiellement les démarches de l'esprit qui cherche à démontrer un théorème simple de géométrie.

Plus ambitieusement, Newell et Shaw employaient une approche analogue à la mise sur pied d'un résolveur général de problème.

Dans la deuxième voie, Gilmore appliquait le théorème de Herbrand pour vérifier diverses formules du calcul des prédicats. Dans les couloirs Prawitz et Voghera distribuaient un résumé de leurs propres recherches dans ce domaine.

Depuis lors chacune de ces deux tendances s'est développée. La première, sous le nom d'Intelligence Artificielle a donné lieu à plus de colloques et de conférences que la seconde.

Il nous a paru équitable de rétablir quelque peu l'équilibre tout en donnant à des représentants de l'Intelligence Artificielle l'occasion de commenter leurs recherches. Il s'agira donc, ici surtout, de la théorie et des pratiques basées sur les méthodes de Herbrand et sur les aspects les plus techniques et les plus profonds de la logique mathématique.

Je m'en excuse auprès d'une partie de l'auditoire mais il y verra la marque que si notre Institut de Recherche se veut appliqué il croit que les bonnes applications ne peuvent surgir que de la réflexion théorique la plus audacieuse et la plus rigoureuse.

N'attendez pas, pour terminer, que je parle de l'avenir. Permettez-moi cependant d'évoquer deux faits:

En 1869, Jevons construisit la première machine pour résoudre le problème de la décision dans le calcul des propositions. Les cas traités par le "piano logique" de Jevons ne sont, de nos jours, que des étapes infinitésimales dans les calculs qui s'effectuent quotidiennement dans tous les Centres de Calculs.

Depuis 1959, la vitesse et la capacité des ordinateurs ont été multipliés par cent et l'efficacité des algorithmes a été amélioré d'autant. Libéré de toute contrainte technologique c'est à vous et à vos élèves que reviendra la charge de faire, non seulement, progresser la théorie dans le domaine de la Démonstration Automatique mais encore d'améliorer son rendement pratique.

Je ne veux pas terminer sans remercier les organisateurs de ce colloque: le Professeur SCHUTZENBERGER, Directeur de Recherche à l'IRIA, et les Professeurs LACOMBE et NOLIN, ainsi que Mademoiselle BRICHETEAU, qui se sont dépensés sans compter pour l'organisation et la réussite de ce colloque.

PRESENTATION D'UN LANGAGE DE FORMALISATION DES DEMONSTRATIONS MATHEMATIQUES NATURELLES

André ARNOLD

1. Introduction

De plus en plus la logique mathématique, dont le but initial est d'exprimer les objets et le raisonnement mathématique dans un langage formel, afin de réduire les démonstrations à un simple calcul, se constitue en branche autonome des mathématiques. Les études des logiciens semblent avoir de moins en moins d'incidences sur le travail habituel du mathématicien. Elles portent davantage sur l'étude des théories et du raisonnement en général que sur l'étude du raisonnement tel qu'il est utilisé dans un quelconque ouvrage de mathématiques. Les systèmes de déduction naturels ont été relativement peu traités.

Dans le domaine de la démonstration automatique, des résultats intéressants ont été obtenus, mais sont difficilement exploitables, entre autres à cause de la difficulté de transcription d'un énoncé en langue naturelle en une formule du calcul des prédicats. D'une part, la démonstration fournie ne ressemble que de très loin à la démonstration au sens où l'entend un mathématicien. Enfin, les processus de démonstration sont encore purement combinatoires et ne font pratiquement pas intervenir les résultats intermédiaires d'une théorie (définitions, lemmes et théorèmes).

Pour améliorer le rendement de la démonstration automatique et pour la rendre plus proche de la démonstration, il serait bon d'y introduire des heuristiques. Pour ce but particulier et aussi pour des raisons d'ordre pédagogiques ou autres, il est intéressant de connaître de façon plus précise ce qu'est une démonstration. Un travail préalable à cette étude est de formaliser les démonstrations, c'est-à-dire de les écrire dans un langage formel dont la sémantique et la syntaxe sont bien définies (en remarquant que dans la logique mathématique, les deux points de vue syntaxique et sémantique sont pratiquement équivalents : une démonstration syntaxiquement correcte l'est aussi sémantiquement).

Un tel langage une fois défini, la vérification d'une démonstration se réduit à un problème d'analyse syntaxique et peut donc être réalisée sur machine.

Paul Abrahams a proposé un système de formalisation qui est en fait une application du langage LISP au langage mathématique.S'il a l'avantage de se rapprocher davantage de la langue naturelle tant au point de vue de l'écriture qu'à celui de la structure d'une démonstration, il reste encore trop fortement lié à la syntaxe de LISP et demande pour être compris une connaissance préalable de LISP et de son écriture fortement parenthésée.

Le langage que nous proposons ici est indépendant de tout langage de programmation et reste assez près de la langue naturelle pour être assimilable par un mathématicien ayant un minimum de connaissances en calcul des prédicats, et en programmation juste ce qu'il faut pour savoir qu'on ne remplace pas impunément un signe par un autre.

Les expressions mathématiques sont écrites dans un formalisme très voisin de celui du calcul des prédicats. Une démonstration est une suite de lignes, chacune étant une définition, un théorème, ou une expression suivie soit d'une "justification" indiquant comment obtenir cette expression à partir des précédentes, soit par une autre démonstration. Nous introduisons ainsi une structure de bloc à l'intérieur d'une démonstration. D'autres structures de blocs sont utilisées pour la quantification et pour la dérivation sous conditions. Ce travail étant une approche d'un domaine encore très peu exploré, il soulève plus de questions qu'il n'en résout. Nous nous sommes efforcés de cerner et de préciser ces questions en suspens en donnant parfois les directions dans lesquelles il nous semble possible de trouver leur solution.

Comme application immédiate de cette formalisation des démonstrations, nous pensons bien sûr à la vérification automatique, mais nous espérons que ce travail pourra servir d'outil dans l'étude de domaines tels que l'heuristique en démonstration automatique, l'enseignement programmé des mathématiques, la simulation du raisonnement en intelligence artificielle.

2. Les expressions mathématiques

2.1 Les objets

Nous nous donnons un alphabet formé de lettres, de chiffres, et de tout autre signe dont nous aurons besoin, à l'exclusion de ceux que nous rencontrerons par la suite qui jouent un rôle particulier.

Les mots formés sur cet alphabet seront appelés identificateurs. D'une manière que nous verrons plus loin, certains identificateurs seront considérés comme des variables.

Une variable est (ou désigne) un objet mathématique considéré comme un tout. Au moyen d'objets et d'autres signes, on peut former de nouveaux objets :

- par composition par un opérateur (un opérateur est un identificateur possédant certaines propriétés particulières qu'on donnera plus loin)

exemple U(A,B)
x + y

- en faisant suivre un objet d'une liste d'objets entre parenthèses

exemple f(x,y)

- en formant des ensembles en extension

exemple {x, y, f(x), f(y)}

ou en compréhension

exemple {x : E(x)} où x est une variable et E(x) une expression mathématique contenant x.

2.2 Les prédicats

Les objets sont reliés entre eux par des prédicats qui sont de deux formes :

- un objet, suivi du symbole de base est un, suivi d'un identificateur de prédicat suivi éventuellement d'une liste d'objets entre crochets.

exemple B est une boule [E, d, r, a]

- deux objets séparés par les symboles de base = ou ∈ ou par un identificateur utilisé comme symbole relationnel

exemple $x = y$
$x \in E$
$x\ r\ x$

2.3 Les expressions

Une expression est soit un prédicat, soit plusieurs prédicats assemblés par des connecteurs et des quantificateurs, comme en logique classique.

Cependant, pour se rapprocher davantage de l'écriture habituelle nous avons donné à ces connecteurs des priorités différentes. Par ordre de priorité décroissante ces connecteurs sont :

la virgule (et) ,

<==> et ==>

ou

∿ (non)

les deux connecteurs <==>et ==> ne sont pas associatifs. On pourra écrire a,b,c
mais pas a ==> b ==> c

Les quantificateurs sont ∀,∃, ∃! (il existe un et un seul). Ils sont suivis d'une variable et mis entre parenthèses. Les variables quantifiées et celles servant à définir un ensemble en compréhension sont dites substituables. Les autres sont libres.

On pourra préciser qu'une variable quantifiée est astreinte à vérifier certaines conditions en écrivant

(Qx : E(x))

où Q est l'un des trois quantificateurs et E(x) une expression contenant x comme variable libre. Des expressions telles que

(∀x : E(x))P(x) (∃x : E(x))P(x) (∃!x : E(x))P(x)

sont respectivement équivalentes à

(∀x)(E(x) ==> P(x)) (∃x)(E(x).P(x)) (∃!x)(E(x).P(x))

exemple d'expression

G est un groupe [o,e] ==> (∀x : x∈G)(∀y : y∈G)(∀z)(xoy = z ==> z∈G)

3. Partie utilitaire

Avant de commencer une démonstration, il faut se donner un certain nombre de renseignements utiles comme des objets propres à la théorie considérée, ses définitions, ses axiomes, des théorèmes supposés déjà démontrés, puis l'énoncé proprement dit du théorème à démontrer.

Pour introduire des objets, on écrit le symbole de base soit suivi d'une liste de variables terminée par un point-virgule. Chaque variable étant éventuellement suivie d'une expression entre parenthèses indiquant les conditions qu'elle doit vérifier.

exemple

soit x(x∈G), y(y∈G), z ;

Ensuite on écrira une liste de définitions. Chaque définition est séparée des autres par un point-virgule. Une définition de prédicat est formée du symbole de base définition suivi d'un prédicat dont les objets sont des variables, suivi du symbole de base = def suivi d'une expression.

exemple définition A est une partie de [B] = def
(∀x)(x∈A ⟹ x∈B)

Une définition d'opérateur a la même forme excepté qu'on remplace est un et l'identificateur de prédicat par le symbole compose

exemple définition U compose [A,B] = def
(∀x)(x∈U(A,B) ⟺ x∈A ou x∈B);

Les axiomes de la théorie sont écrits sous forme d'une expression qui est précédée du symbole de base h.

Chaque théorème composant la liste de théorèmes est formé du symbole de base théorème suivi d'un identificateur de théorème suivi éventuellement d'une liste de variables entre crochets suivi d'une hypothèse (éventuellement) et d'une conclusion. Ces hypothèse et conclusion sont des expressions précédées respectivement de h et c

exemple théorème EE[A,B] h
(∀x)(x∈A ⟺ x∈B) c A = B ;

L'énoncé du théorème à démontrer est formé d'une expression qui peut éventuellement être précédée d'une introduction de variables et de nouvelles hypothèses

exemple soit E,d ;
----------;
soit r, a, B ; h B est une boule [E, d, r, a] ;
B est un ouvert [E,d]

Toutes les expressions qui figurent dans cette partie utilitaire, ainsi que celles qui apparaîtront par la suite dans la démonstration doivent vérifier certaines conditions sur les identificateurs qu'elles contiennent : toute variable libre qui n'est pas paramètre d'une définition ou d'un théorème doit être introduite ;

tout identificateur de prédicat doit figurer dans la partie gauche d'une définition. Ce sont les mêmes conditions, mutatis mutandi, qu'on a en Algol sur les déclarations d'identificateurs et de procédures.

Un "texte" complet dans le langage que nous définissons sera constitué d'une introduction (éventuellement vide), d'une liste de définitions (éventuellement vide), des axiomes (éventuellement vide), d'une liste de théorèmes, et de l'énoncé du théorème à démontrer suivi de sa démonstration.

4. Démonstration

L'expression à démontrer qui termine l'énoncé est suivie d'une démonstration. Une démonstration peut prendre trois formes désignées par a_1, a_2, a_3.

Une démonstration du type a_1 (démonstration directe) est formée d'une suite de lignes encadrée par les symboles en effet et cqfd.

Une ligne est

- une définition ou un théorème

ou- une étiquette suivie de deux-points suivi d'une expression, suivie d'une preuve

ou- une étiquette suivie de deux-points suivi de l'un des trois blocs b_1, b_2 ou b_3.

Une preuve est soit une autre démonstration, soit une justification précédée du symbole par (on parlera au paragraphe suivant des justifications).

Une démonstration du type a_1 est correcte si l'expression figurant dans la dernière ligne de la démonstration est identique (à une substitution près des variables substituables) à l'expression à démontrer ;

après quoi on peut ignorer tout ce qui figurait dans la démonstration.

Une démonstration correcte du type a_2 (démonstration par l'absurde) a l'allure suivante :

A <u>absurde</u> n:∼A;

...
k:B (ou∼B) <preuve> ;
...
m:∼B (ou B) <preuve> <u>contradiction avec</u> k

Une démonstration correcte du type a_3 (démonstration par récurrence) s'écrit, sous réserve que l'on ait défini ≥ et +)

(∀m : m≥n)P(m) <u>récurrence</u>

α : P(n) <preuve> ;
β : <u>soit</u> m(m≥n) ; γ : <u>h</u> P(m) ;
δ : P(m+1) <preuve> <u>ok</u>

La structure récursive des démonstrations, analogue à celle des blocs en Algol, outre son intérêt pratique, permet de mieux mettre en évidence le cheminement de la démonstration.

Les trois blocs b_1, b_2 et b_3 sont utilisés pour appliquer respectivement la déduction sous conditions, la quantification universelle et existentielle. De façon plus précise le bloc b_1 commence par le symbole <u>h</u> suivi d'une expression qu'on suppose vraie, puis une liste de lignes, puis une dernière ligne formée d'une étiquette, d'une expression et du symbole <u>déduction</u>. Un bloc b_1 correctement écrit a donc la forme

<u>h</u> f_o ;
----
e_n : f_n <preuve> ;
e_{n+1} : f_{n+1} <u>déduction</u>

où f_{n+1} est identique à $\bar{f}_o \Longrightarrow \bar{f}_n$ avec $\bar{f}$ défini de telle façon que l'expression $\bar{f}_o \Longrightarrow \bar{f}_n$ soit correctement parenthésée, et sans parenthèses superflues.

Le bloc b_2 a la forme

<u>soit</u> x_1, x_2 --- x_p ;

----

e_n : f_n <preuve> ;

e_{n+1} : f_{n+1} <u>généralisation - ∀</u>

où x_1, x_2 --- x_p sont des variables non encore introduites et où f_{n+1} est identique à $(\forall x_1)(\forall x_2)$ --- $(\forall x_p)\ f_n^*$, avec $f^* = f$ si f est un prédicat ou une expression déjà quantifiée, $f^* = (f)$ sinon

<u>Remarque</u> Si une variable introduite est suivie d'une expression la variable quantifiée est suivie de la même expression.

<u>exemple</u> <u>soit</u> x(x∈G) ... ;

.....

e : (∀x : x∈G) ...

Un bloc b_3 peut commencer de trois façons différentes :

. <u>prenons</u> x <u>défini par</u> E(x) <u>existence</u> $(\exists x)E^*(x)$ <preuve> ;

où E(x) est une expression contenant x ;

. <u>prenons</u> x <u>déjà défini</u> ;

où x est une variable déjà introduite

. <u>prenons</u> x = X ;

où X est un objet,

et ce bloc se termine par

e_n : f_n <preuve> ;

e_{n+1} : f_{n+1} <u>généralisation - ∃</u>

avec f_{n+1} identique à $(\exists x)f_n^*$.

Ces trois blocs, dont la signification logique est claire, permettent d'exprimer commodément les règles correspondantes. Pour des raisons bien évidentes, toutes les expressions qui figurent dans un tel bloc, à l'exception de la dernière, et les variables introduites en tête de ce bloc, doivent être ignorées à l'extérieur du bloc.

Justifications

La justification qui suit une expression indique comment on a obtenu cette expression à partir des expressions précédentes repérées par leur étiquette.

Etant donné une étiquette, nous appellerons expression étiquetée l'expression à laquelle renvoie cette étiquette. On retrouve l'expression étiquetée à partir de l'étiquette de la manière suivante :

- si l'étiquette est suivie d'une expression (on ne tient pas compte des deux points qui doivent toujours suivre une étiquette) c'est cette expression qui est l'expression étiquetée.
- si l'étiquette est suivie d'une hypothèse, l'expression étiquetée sera l'expression qui suit le symbole h
- si l'étiquette est suivie de prenons <variable> défini par <expression,> l'expression étiquetée est celle qui suit le symbole défini par
- si l'étiquette est suivie de prenons <variable> = <objet>, l'expression étiquetée est <variable> = <objet>.
- les hypothèses figurant dans la partie utilitaire et dans l'énoncé ne sont pas précédées d'étiquette. On les repérera par des étiquettes conventionnelles, par exemple 0 et 00. Si on rencontre l'étiquette 0 (respectivement 00) l'expression étiquetée sera l'hypothèse de la partie utilitaire. (Respectivement de l'énoncé).

- l'expression suivant une variable introduite sera repérée par l'étiquette de la ligne d'introduction suivie de deux points suivi de la variable.

<u>exemple</u> 4 : <u>soit</u> x(x∈E), y, z;

x∈E est l'expression étiquetée par 4 :x.

Nous donnons ici une liste de justifications. Cette liste n'est ni exhaustive, ni définitive. A l'usage il s'avèrera peut-être que certaines justifications sont inutiles d'autres mal définies alors qu'il serait intéressant de pouvoir en utiliser d'autres non signalées. Il ne sera pas difficile de modifier cette liste en conséquence.

Aprés chaque justification nous indiquons comment on obtient l'expression justifiée à partir des expressions étiquetées.

☆ <u>modus-ponens</u> e_1, e_2

$$\left.\begin{array}{l} e_1 : f \\ e_2 : \bar{f} \Longrightarrow \bar{g} \end{array}\right\} \quad g$$

☆ <u>conjonction de</u> $e_1, e_2, \ldots, e_n$

$$\left.\begin{array}{l} e_1 : f_1 \\ \ldots \\ e_n : f_n \end{array}\right\} \quad f_1, f_2, \ldots, f_n$$

☆ <u>cas</u> $i_1, i_2, \ldots, i_n$ <u>de</u> e

$$\left. e : f_1, f_2, \ldots, f_p \right\} \quad f_{i_1}, f_{i_2}, \ldots, f_{i_n}$$

☆ <u>composition de</u> Θ <u>par</u> $v_1, v_2, \ldots, v_n$ (v_i est un objet)

$$\left.\begin{array}{l} \underline{\text{définition}}\ \Theta\ \underline{\text{compose}}\ [x_1, x_2, \ldots, x_n] \\ = \text{def } E(\Theta, x_1, x_2, \ldots, x_n); \end{array}\right\} \quad E(\Theta, v_1, v_2, \ldots, v_n)$$

* <u>définition de</u> e

<u>définition</u> X <u>est un</u> chose [Y,Z]
= def E(X,Y,Z);
e : A <u>est un</u> chose [B,C]
(A,B,C sont des objets)
} E(A,B,C)

* <u>définition en</u> $e_1, e_2, \ldots, e_n$

<u>définition</u> X <u>est un</u> chose [Y,Z]
= def E(X,Y,Z)
$e_1 : f_1$
$e_2 : f_2$
...
$e_n : f_n$
} (avec $E(A,B,C) = f_1, f_2, \ldots, f_n$)
} A <u>est un</u> chose [B,C]

* <u>application du théorème</u> machin [x,y] $e_1, e_2, \ldots, e_n$ (x,y objets)

<u>théorème</u> machin [u,v] <u>h</u> E(u,v)
<u>c</u> F(u,v)

$e_1 : f_1$
...
$e_n : f_n$
(avec $E(x,y) = f_1, f_2, \ldots, f_n$)
} F(x,y)

* <u>décomposition de</u> e

e : a_1 <u>ou</u> a_2 <u>ou</u> .... <u>ou</u> a_n } a_1 ==> b,..., a_n ==> b

* <u>recomposition de</u> e

(la même chose en sens inverse)

* <u>implication 1 en</u> e

e : f <==> g } f ==> g

* <u>implication 2 en</u> e

e : f <==> g } g ==> f

* <u>double implication en</u> e

e : f ==> g, g ==> f } f <==> g

* <u>particularisation de</u> e <u>par</u> $v_1, v_2, \ldots, v_n$ (les v_i sont des objets).

e : $(\forall x_1) \ldots (\forall x_p) E(x_1, \ldots, x_p)$

} $(\forall x_{n+1}) \ldots (\forall x_p) E(v_1, v_2, \ldots, v_n, x_{n+1}, \ldots, x_p)$

Dans le cas où la variable quantifiée est suivie d'une expression, l'objet qu'on lui substitue doit vérifier la même expression, ce dont on s'assurera en mettant derrière l'objet l'étiquette de cette expression. On pourra cependant s'en dispenser si cette expression figure dans l'introduction de l'objet en question.

<u>exemple</u> 2 : <u>soit</u> v(v∈E) ;
3 : w∈E <preuve> ;
4 : (∀x:x∈E)(∀y:y∈E)P(x,y) <preuve> ;
5 : P(v,w) <u>par particularisation de</u> 4 <u>par</u> v,w : 3 ;

* <u>simplification de</u> e

sert à enlever des quantificateurs superflus

* <u>transitivité de l'implication</u> e_1, e_2

e_1 : f ==> g
e_2 : g ==> h } f ==> h

* <u>égalité</u> e_1 <u>dans</u> e_2

e_1 : X = Y
e_2 : P(X,X,Z) } P(X,Y,Z)

* <u>identité</u> e_1 <u>dans</u> e_2

au lieu de substituer des objets égaux, on substitue des expressions logiquement équivalentes.

* contraposition de e

$e : a \Longrightarrow b$ $\Big\}$ $\tilde{b} \Longrightarrow \tilde{a}$ (où $\tilde{f}$ est la négation de f, après simplification d'écriture)

* appartenance à l'ensemble e_1, e_2

$e_1 : Y = \{y : P(y)\}$
$e_2 : x \in Y$
$\Big\}$ $P(x)$

* définition de l'ensemble e_1, e_2

$e_1 : Y = \{y : P(y)\}$
$e_2 : P(x)$
$\Big\}$ $x \in Y$

* formation de l'ensemble e

$e : (\forall y)(y \in X \Longleftrightarrow P(y))$ $\Big\}$ $X = \{y : P(y)\}$

* unicité en e_1, e_2, e_3

$e_1 : (\exists ! x)P(x)$
$e_2 : P(a)$
$e_3 : P(b)$
$\Big\}$ $a = b$

* existence unique en e_1, e_2

$e_1 : (\exists x)P(x)$
$e_2 : (\forall x)(\forall y)((P(x), P(y)) \Longrightarrow x = y)$
$\Big\}$ $(\exists ! x)P(x)$

D'autres justifications, dont le fonctionnement n'est pas encore complètement précisé, permettent de faire d'autres raisonnements.

6. Autres possibilités du langage

6.1 Générateurs d'expressions.

Lorsqu'on appliquera successivement plusieurs justifications, on pourra les regrouper dans une même ligne et prendre comme expression justifiée la dernière expression obtenue, de la même manière qu'on compose des fonctions : au lieu d'écrire
$y = h(x)$, $z = g(u)$, $t = f(y,z)$, on peut écrire directement $t = f(h(x),g(u))$.

En effet, la plupart des justifications peuvent être considérées comme des applications univoques de $\mathcal{E}^*$ dans $\mathcal{E}$, où $\mathcal{E}$ est l'ensemble des expressions et $\mathcal{E}^* = \emptyset \cup \mathcal{E} \cup \mathcal{E}^2 \ldots \cup \mathcal{E}^n \ldots$ On pourra donc composer les justifications comme on compose les applications.

exemple au lieu de :

3 : x≤y => x∈H, x≤y, y∈H <preuve> ;
4 : x≤y => x∈H par cas 1 de 3 ;
5 : x≤y par cas 2 de 3 ;
6 : x∈H par modus-ponens 5, 4 ;

nous écrirons :

3 : x≤y => x∈H, x≤y, y∈H <preuve> ;
4 : x∈H par modus-ponens (car 2 de 3), (cas de 1 de 3) ;

Cette transformation se fait en remplaçant les étiquettes par les justifications qui ont permis d'obtenir les expressions étiquetées. On met cette justification entre parenthèses pour éviter les ambiguités (exemple : .... conjonction de 2, définition en 3,4 peut se comprendre conjonction de 2, (définition en 3,4) ou conjonction de 2, (définition en 3), 4).

Dans la définition des justifications, nous remplacerons donc partout <étiquette> par <générateur d'expression> et dans l'explication de leur fonctionnement nous ne nous réfèrerons plus aux expressions étiquetées mais aux expressions générées (ce qui ne change évidemment rien dans le cas où le générateur d'expression est une étiquette).

L'utilisation des justifications composées permet donc de réduire le nombre de lignes d'une démonstration en n'y faisant pas figurer certaines expressions intermédiaires,

6.2 Génération de théorèmes

Lorsqu'on vient de démontrer une expression, on pourra considérer cette expression comme la conclusion d'un théorème dont les paramètres et l'hypothèse sont donnés par l'énoncé. Le symbole de base conserver permettra de donner un nom à ce théorème et de le ranger dans la liste de théorèmes. Ceci permettra de ne démontrer qu'une seule fois plusieurs expressions qui ne diffèrent que par les objets qui y figurent.

. Questions ouvertes

Dans son état actuel, ce langage ne permet pas encore de formaliser n'importe quelle démonstration mathématique.

Des difficultés apparaissent pour formaliser les variables numériques, en particulier lorsqu'on les utilise comme indices :

$$x_{i+j} = f(x_i, x_j)$$

ou dans des raisonnements par récurrence ou par induction. Lorsque la valeur de ces variables numériques n'est pas connue explicitement, on fait généralement apparaître des points de suspension dans la démonstration. Il faudrait définir avec précision la signification et les conditions d'emploi de ces points de suspension.

De manière plus générale, il n'est pas possible de traiter actuellement des expressions du second ordre (on peut substituer des objets mais pas des expressions) à moins d'utiliser certaines astuces comme remplacer une expression par son extension (ensemble des objets pour lesquels elle est vraie).

D'autre part les démonstrations formalisées sont encore très longues, par rapport à ce qu'écrivait un mathématicien, parce qu'il faut préciser tous les pas du cheminement logique. Pour se rapprocher davantage de la démonstration naturelle, il faudrait pouvoir sauter certaines étapes. Pour cela on peut considérer que certaines dérivations sont "évidentes" et ne pas les signaler. Il faudrait aussi pouvoir créer des "procédures de raisonnement" analogues aux procédures d'un langage de programmation, et qui correspondraient à des phrases telles que : "De la même façon on montre que....."

Pour résoudre ce dernier problème on peut envisager la solution suivante que nous ne donnons qu'à titre indicatif car elle n'a pas encore été expérimentée.

Soit une expression suivie d'une démonstration qu'on veut conserver après l'avoir vérifiée. On mettra alors une parenthèse fermée après l'étiquette de cette expression. Lorsqu'une autre expression se démontrera de la même façon, au lieu de réécrire la démonstration, on se contentera d'indiquer ce qui la différencie de la démonstration conservée, ces différences pouvant porter sur des objets, des étiquettes et même des expressions. On écrira donc quelque chose comme :

$e_1)$: E <démonstration> ;

e_2 : F <u>comme pour</u> e_1 <u>avec</u> x_1 <u>remplacé par</u> $y_1, \ldots, x_n$ <u>remplacé par</u> y_n, l_1 <u>remplacé par</u> $l'_1, \ldots, l_p$ <u>remplacé par</u> l'_p <u>et</u> $q_1 : A_1$ <u>et</u> ... <u>et</u> $q_m : A_m$;

(les x_i et y_i sont des objets, les l_i et l'_i des étiquettes figurant à l'extérieur de la démonstration conservée, les q_i des étiquettes figurant à l'intérieur, les A_i des expressions).

Un symbole particulier devrait permettre de signaler qu'on n'aura plus besoin d'une démonstration conservée et qu'on peut donc l'oublier.

Enfin, il serait intéressant de pouvoir utiliser le "contexte" dans une démonstration, ce qui permettrait d'omettre certaines précisions qu'on sous-entend également dans une démonstration normale, en particulier de ne pas préciser certains arguments de certains prédicats, etc.

Applications

L'application la plus immédiate est la vérification automatique de démonstrations écrites dans ce langage, ce qui permet également de déceler les avantages et les inconvénients du langage dans son état actuel. Un tel travail a été réalisé par MM. Henneron et Guillerminet sur l'IBM 7044 de Grenoble pour une partie du langage que nous venons de définir. La plupart des possibilités récemment introduites restent encore à tester.

Il serait aussi souhaitable d'étudier de façon plus approfondie la compatibilité et la complétude de l'ensemble des "justifications" que nous utilisons.

Une fois les démonstrations formalisées au moyen de ce langage, on disposera d'un corpus permettant d'étudier expérimentalement les démonstrations telles qu'elles figurent dans la littérature mathématique, ce qui pourrait conduire à des études intéressantes dans des domaines tels que l'intelligence artificielle et les mécanismes de l'intuition, la pédagogie des mathématiques, etc...

Pour terminer, je tiens à exprimer ma reconnaissance à M. Kuntzmann qui m'a donné l'idée de ce travail, et à tous ceux qui, par l'intérêt qu'ils y ont porté, m'ont encouragé à le poursuivre.

Annexe

Exemple de démonstration formalisée.

remarques - l'introduction de symboles relationnels comme $\subset$ n'est pas encore bien définie mais ne semble pas devoir poser de difficultés.

- le prédicat "est un voisinage" est un prédicat de base de la théorie considérée, c'est pourquoi sa définition ne contient pas d'expression définissante.

L'ensemble des points d'un espace topologique est un ouvert.

Soit $\subset$; définition x est un voisinage ;
définition x est un point = def($\exists$F) (F est un voisinage, x$\in$F) ;
définition F est un voisinage de [x] = def F est un voisinage, x$\in$F ;
définition F est une classe de points = def($\forall$z) (z$\in$F => z est un point) ;
définition x est un point intérieur à [F] = def F est une classe de points, ($\exists$G) (G est un voisinage de [x], G$\subset$ F) ;

définition F est un ouvert = def F est une classe de points, ($\forall$x) (x$\in$F => x est un point intérieur à [F]) ;
définition $\cap$ compose [A,B] = def($\forall$x) (x$\in$ $\cap$(A,B) <=> (x$\in$A, x$\in$B)) ;

h ($\forall$x)($\forall F_1$) ($\forall F_2$) ((F_1 est un voisinage de [x], F_2 est un voisinage de [x]) => (($\exists$G) (G $\subset$ $\cap$(F_1,F_2),

G est un voisinage de [x])),

(∀x)(∀y) ((x est un point, y est un point, ∼x = y) =>

(∃F)(∃G)(F est un voisinage de [x], G est un voisinage de [y],

∩ (F,G) = Ø)) ;

théorème de l'ensemble vide [E] c E = Ø <=> ∼(∃x) x∈E ;

théorème définition de l'inclusion [A,B] c A ⊂ B <=>

(∀x)(x∈A => x∈B) ;

Soit P ; h P = {x : x est un point} ;

P est un ouvert

en effet

1 : P est une classe de points

en effet

11 : Soit x ;

12 : h x∈P ;

13 : x est un point par appartenance à l'ensemble 00,12 ;

14 : x∈P => x est un point déduction ;

15 : (∀x)(x∈P => x est un point) généralisation-∀ ;

16 : P est une classe de points par définition en 15

cqfd ;

2 : (∀x)(x∈P => x est un point intérieur à [P])

en effet

21 : Soit x ;

22 : h x∈P ;

23 : x est un point intérieur à [P]

en effet ;

231 : Prenons F défini par F est un voisinage,

x∈F existence (∃F) (F est un voisinage, x∈F)

en effet 2310 : x est un point par appartenance

à l'ensemble 00,22 ;

2311 : (∃F)(F est un voisinage, x∈F) par cas 1

de (définition de 2310) cqfd ;

232 : F est un voisinage de [x] par définition en 231 ;

233 : F⊂P

en effet

2331 : Soit z ;

2332 : h z∈F ;

2333 : prenons F déjà défini ;

2334 : F est un voisinage, z∈F par conjonction de (cas 1 de 231), 2332 ;

2335 : (∃F) (F est un voisinage, z∈F) généralisation-∃ ;

2336 : z est un point par définition en 2335 ;

2337 : z∈P par définition de l'ensemble 00,2336

2338 : z∈F => z∈P déduction ;

2339 : (∀z) (z∈F => z∈P) généralisation-∀ ;

2340 : F ⊂ P <=> (∀x) (x∈F <=> x∈P) par application du théorème définition de l'inclusion [F,P] ;

2341 : (∀x) (x∈F => x∈P) => F ⊂ P par implication 2 en 2340 ;

2342 : F⊂P par modus-ponens 2239,2341

cqfd ;

234 : F est un voisinage de [x], F⊂P par conjonction de 232,233 ;

235 : (∃F) (F est un voisinage de [x], F⊂P) généralisation-∃ ;

236 : P est une classe de points, (∃F) (F est un voisinage de [x], F ⊂ P) par conjonction 1, 235 ;

237 : x est un point intérieur à [P] par définition en 236

cqfd ;

238 : x∈P => x est un point intérieur à [P] déduction ;

239 : (∀x) (x∈P => x est un point intérieur à [P])
généralisation-∀

cqfd ;

3 : P est un ouvert par définition en (conjonction de 1,2) ;

cqfd

Références

ARNOLD, A. Formalisation des démonstrations mathématiques

Thèse Lille (1968)

HENNERON, GUILLERMINET

Application d'un langage de formalisation des démonstrations mathématiques

Rapport, Institut Polytechnique de Grenoble (1968)

The mathematical language AUTOMATH, its usage, and some of its extensions

N.G. de Bruijn

1. Introduction.

1.1 AUTOMATH is a language which we claim to be suitable for expressing very large parts of mathematics, in such a way that the correctness of the mathematical contents is guaranteed as long as the rules of grammar are obeyed.

Since the notions "mathematics" and "expressing" are rather vague, we had better discuss a specific example. Assume we have a very elaborate textbook on complex function theory presenting everything from scratch. That is, we start with chapters on logic and inference rules, set theory, the number systems, some geometry, some topology, some algebra, and we never use anything that is not derived, unless it has been explicitly stated as an axiom. Assume the book has been most meticuously written, without leaving a single gap. Then we claim it is possible to translate this text line by line into AUTOMATH. The grammatical correctness of this new text can be checked by a computer, and that can be considered as a final complete check of the given piece of mathematics. Moreover we claim that it is possible to do so in practice. The line by line translation will be a matter of routine; the main difficulty lies in the detailed presentation of such a large piece of mathematics. The mere labour involved in the translation will not increase if we proceed further into mathematics.

1.2 AUTOMATH was developed in 1967-1968 at the Technological University, Eindhoven, The Netherlands. The author is indebted to Mr. L.S. van Benthem Jutting for very valuable help in trying out the language in several parts of mathematics, and both to him and to Mr. L.G.F.C. van Bree for their assistance with the programming (in ALGOL) of processors by means of which books written in AUTOMATH can be checked. In particular, Mr. Jutting is currently translating Landau's "Grundlagen der Analysis".

1.3 In this paper we shall not attempt a complete formal definition of AUTOMATH, for which we refer to the report "AUTOMATH, a language for mathematics" by N.G. de Bruijn, Report 68-WSK-05, Technological University Eindhoven, The Netherlands. Nevertheless we hope to make the language intuitively clear in this paper. After all, the author feels that very little is essentially new in AUTOMATH, that it is very close to the way mathematicians have always been writing, and that the abbreviating system used in AUTOMATH has been taken from existing mathematical habits. The way we handle propositions and assertions will be novel among other things.

1.4 One of the principles of the language is that the reader (be it a human being or a computer) never has to search in the previous text for definitions or arguments. The text presented to him tells him precisely where to find information needed for checking that text.

1.5 We indicate the possibility of building languages defined in terms of AUTOMATH but adapted to special purposes (superimposed languages, see sec. 10). This is one of the reasons for keeping AUTOMATH as primitive as possible. Actually it is little more than what might be called <u>the art of substitution</u>. AUTOMATH has an even more primitive sub-language PAL (see sec. 4), but PAL is definitely too primitive to deal with things like predicates, quantifiers and functions. As a preliminary, we shall introduce a simple language SEMIPAL, which is not a sublanguage of PAL.

1.6 An AUTOMATH book is a sequence of lines written according to the rules of grammar. An important feature is that things which have been derived in a book (e.g. inference rules, definitions, theorems) can be applied later in that same book. It turns out to be possible that even very primitive parts of mathematical logic can be explained in that book, and therefore it is unnecessary to feed that kind of logic into the grammar.

1.7 There is one vital thing that we do not attempt to formalize: the interpretation. When reading or writing a book in a formal language like AUTOMATH, we try to be constantly aware of the relation between the text and the (mathematical or non-mathematical) objects we imagine that the text refers to. It is in this sense that many words occurring in the book (identifiers) are names of the objects outside. The book itself deals with names only. There may be several different interpretations, and there seems to be no way to discuss these interpretations in the book.

2. Preliminary description of the language.

2.1 An AUTOMATH book is written in lines. Everything we say is said in a certain context; we shall attach a context indicator (or indicator for short) to every line. Usually the context structure can be described by a set of nested blocks (see 3.10), such as in a system of natural deduction. Lines written in a block have a kind of validity inside that block.

The context structure will make it possible to express a certain functional relationship. On top of that we have another way of dealing with functions: something that is essentially Church's lambda conversion calculus. Although these two features do not make each other entirely superfluous, they create a certain abundancy in the language. By virtue of this abundancy, many things can be written in various ways. One might experience this as a drawback, but, on the other hand, it gives something of the flexibility of everyday mathematical language.

2.2 In every line a new name (an <u>identifier</u>) is introduced. It is very essential that to every identifier a <u>category</u> is attached. In every-day language this amounts to stating what kind of a thing we are talking about. For example, we might introduce the identifier "two" and say that its category is "integer". We shall not admit that "two" has several categories simultaneously. This may have the drawback that we have to invent different notations for the integer 2 and the complex number 2. Accordingly, we have to express ourselves by means of one-to-one mappings of the integers into the complex numbers, instead of care-free identification. (We should not forget that care-free identification is a matter of tradition. The average mathematician is not inclined to identify a unit matrix with the number 1, but he identifies all 1's he knows as long as they belong to one of the "number systems").

In connection with the above example we remark that it is by no means necessary to write mathematics in such a way that "two" has the category "integer". Another possibility, as well rooted in existing habits as the previous one, is to write that both "two" and "integer" have the category "object", and to add that "two $\in$ integer" is a true statement. If we do this, there is no harm in saying that "two $\in$ complex number" is also true.

2.3 It will be possible to introduce new categories. For this purpose we use the special symbol <u>type</u>. For example, we may introduce the identifier "integer" and attach the category <u>type</u> to it. This will have the consequence that later in the text (at least in the context where "integer" was introduced) we have the right to use "integer" as the category of an identifier.

2.4 Another feature of AUTOMATH is an abbreviation system which is essentially taken from existing conventions in mathematics; this can make the labour of writing and reading bearable, especially if we select suggestive identifiers for all notions introduced in the book. In essence, this abbreviation system occurs already in SEMIPAL.

3. <u>Structure of the lines.</u>

3.1 A line consists of 4 parts:

(i) an indicator,
(ii) an identifier,
(iii) a definition,
(iv) a category.

3.2 In every line the identifier part (ii) is a symbol that has not been used in previous lines. (This stipulation is unusual in every-day mathematics: a symbol like x is used repeatedly in different senses. But assuming we have infinitely many symbols available, it would do no harm to replace all these x's by different symbols whenever necessary.)

An identifier used as identifier part of a line will be called a proper identifier. There is a second kind of identifiers: those that play the rôle of bound variables. Again, in contrast with existing habits we shall use each bound variable only once, and a bound variable has to be different from previously introduced proper identifiers.

There are three kinds of proper identifiers: block openers, primitive notions, and compound notions. This depends on the definition part of the line. If the definition part is —, then the identifier part is called a block opener (or "free variable"). If the definition part is PN, then the identifier part is called a primitive notion. If the definition part is an expression (see sec. 3.3), then the identifier part is called a compound notion.

There is a second classification of identifiers, which bears no relation to the classification above. Some identifiers are object names, others are types. An identifier is a type if and only if it is the identifier part of a line whose category is type. All other identifiers (including bound variables) are called object names.

3.3 The definition part of a line is either an expression or one of the symbols PN or —. If the definition part is an expression, that expression is composed of (i) proper identifiers of previous lines; (ii) bound variables; (iii) the symbols

( ) { } [ ]

which are used as separation marks.

3.4 The category part of a line is either the symbol type or an expression. If it is an expression, we can say the same things as in 3.3.

3.5 The indicator part of a line is either the symbol 0 or a block opener introduced in a previous line.

The indicator is used in order to describe context.

3.6 A book is organized as a string of lines, but the context indicators induce a second structure in the form of a rooted oriented tree. The root is the symbol 0, the other vertices are the identifiers of the lines of the book. The edges are all oriented towards the root. The edge starting at the identifier x points to the indicator of the line that has x as its identifier part.

3.7 As an example we take the following book:

indicator	identifier	definition	category
0	a	PN	type
0	x	—	....
0	b	....	....
x	c	PN	type
x	y	—	type
y	z	—	....
y	d	....	....
0	e	....	....
x	w	—	....
w	f	....	type
y	g	....	....

In this example we have written .... in order to suppress expressions we do not intend to discuss at this moment. (So "...." is not a symbol used in AUTOMATH, but in our discussion about AUTOMATH.) In this example x,y,z,w are block openers, a and c are primitive notions, b,d,e,f,g are compound notions.

The tree of this book is

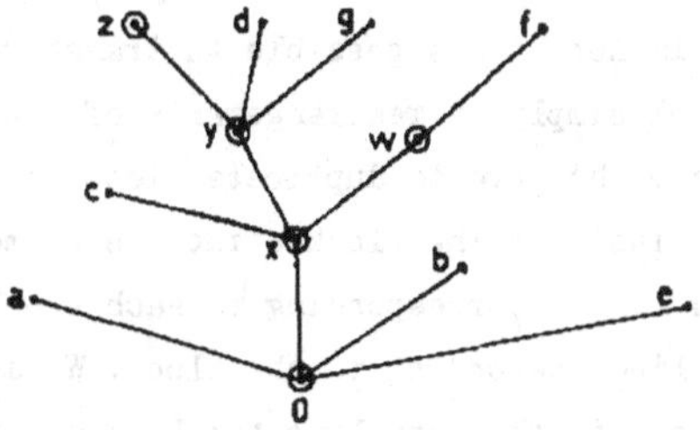

3.8 It has to be remarked that the tree is a combinatorial thing, and that the way it is drawn in a plane is quite irrelevant.

Note that the primitive notions and the compound notions are end-points of the tree. The block openers are usually not end-points.

To every point $\neq 0$ of the tree we can attach the definition part and the category part of the line of which that point is the identifier part. If we do this, the tree contains all the information of the book, and can be referred to as the <u>tree of knowledge</u>. But one thing the tree does not reveal: it does not show the order of the lines in the book. If we want to know whether the tree is grammatically correct, it is useful to know the order of the lines. Given the set of lines of a valid book, there may be several ways to arrange them. The only condition an arrangement has to satisfy is that no expression occurring in a line contains identifiers of later lines. All such arrangements produce legitimate books.

Anyway, if we want to extend the book by a further line then the order of the previous lines is irrelevant. At that moment, it is only the tree of knowledge that counts.

3.9 If p is a point of an oriented rooted tree, different from the root, then we can consider the subtree of all those points of the tree for which the oriented path to the root passes through p (p itself is the root of the subtree). In the case of our tree of knowledge, we shall refer to these subtrees as <u>blocks</u>. In the tree of 3.7, the point x determines the block containing x,c,y,z,d,w,f,g; the block opener of that book is x.

3.10 Quite often a book has been written in an order that makes the block structures immediately clear. This is the case if every block consists of a set of consecutive lines. In this case we shall say we have a <u>nested book</u>. (We remark that it is not always possible to transform a correct book into a correct nested book simply by rearrangements of the lines. In order to get a nested book we might have to duplicate pieces of the text.)

In a nested book we can indicate the block structure by means of vertical bars in front of the lines. Corresponding to each block we draw a vertical line spanning all lines belonging to the block. We agree that if block B is contained in block A, then the line for B is drawn to the right of the line for A.

Once the lines have been drawn, the indicators can be omitted since they can be retraced. In the example below we present a nested book twice, once with indicators, once with bars. The version with the bars is certainly more readable for the human mathematician. A computer will of course prefer the one with the indicators.

0	a	PN	<u>type</u>
0	x	—	<u>type</u>
x	y	—	....
y	b	....	....
x	z	—	....
z	c	PN	....
z	w	—	....
w	d	....	....

```
a      :=   PN     type
|x     :=   —      type
||y    :=   —      ....
||b    :=   ....   ....
||z    :=   —      ....
||c    :=   PN     ....
|||w   :=   —      ....
|||d   :=   ....   ....
```

As in this example we shall always separate identifier part and definition part by the symbol := which suggests that the identifier on the left is defined by the expression on the right.

Needless to say, the vertical bars and the symbol := do not belong to the language. They are just devices for easier reading. Quite often we shall print both the vertical bars and the indicators.

3.11 Sometimes we shall talk about the <u>indicator string</u> of a line. If the indicator is 0, the indicator string is empty. In all other cases the indicator string describes the reversed path from the indicator in question to the root of the tree (excluding the root). For example, the indicator string of the last line in the example of 3.7 is (x,y), the one of the last line in the example of 3.10 is (x,z,w).

4. <u>How to write PAL.</u>

4.1 PAL is a sublanguage of AUTOMATH, in the sense that every correct PAL book is also a correct AUTOMATH book. PAL is quite easy to learn. In PAL we do not use the lambda conversion, and we have no bound variables.

Let us take an example. At this stage the reader must not expect an example with deep mathematical significance, since that would require quite a long book. The interpretation we have in mind is this one: Assume that nat (natural number) and real (real number) are available as categories. If a and b are given reals, then their product is introduced as a primitive notion. If n is a natural number and x is a real, than the power x^n is introduced as a primitive notion. If n is a natural number and y is a real number, then we define $d(y) := y^n$; $e(y) := d(y) \cdot y \ (= y^{n+1})$; $f(y) := d(y) \cdot d(y) \ (= y^{2n})$; $g(y) := e(d(y)) (= y^{n(n+1)})$. In PAL this can be written as follows:

(indicator)	(identifier)		(definition)	(category)	
0	nat	:=	PN	<u>type</u>	1
0	real	:=	PN	<u>type</u>	2
0	a	:=	—	real	3
a	b	:=	—	real	4
b	prod	:=	PN	real	5
0	n	:=	—	nat	6
n	x	:=	—	real	7
x	power	:=	PN	real	8
n	y	:=	—	real	9
y	d	:=	power(n,y)	real	10
y	e	:=	prod(d,y)	real	11
y	f	:=	prod(d,d)	real	12
y	g	:=	e(d)	real	13

This happens to be a nested book in the sense of 3.10, but that does not have any consequence for the present discussion. It is also a very simple case in the sense that the categories are all very simple.

Although we are not going to do it in this paper, it may help the reader to provide the identifier parts (as far as they are not block openers) with the indicator strings in parentheses. That means that he writes prod(a,b) in line 5, power(n,x) in line 8, d(n,y) in line 10, e(n,y) in line 11, f(n,y) in line 12, g(n,y) in line 13. This makes it easy to see what we intend with the other expressions: prod(d,d) indicates that both a and b in prod(a,b) are replaced by d. Now what does e(d) mean in line 13? By line 11, e depends on two variables (n and y). We agree that we add the letters of the indicator string of line 11 on the left until we have enough entries. So e(d) has to be interpreted as e(n,d): the first entry of the string n,y is added on the left. In general: if p is introduced with indicator string $(x_1,\ldots,x_n)$, and if $k < n$, then $p(Z_1,\ldots,Z_k)$ has to be interpreted as $p(x_1,\ldots,x_{n-k}, Z_1,\ldots,Z_k)$.

4.2 Before we describe the rules of PAL, we first describe a simpler language to be called SEMIPAL. This language is different from PAL and AUTOMATH in that it does not attach a category to a line. Its relation to PAL is simple. If we just cancel from a correct PAL book the entire category column, then we get a correct SEMIPAL book. (On the other hand, we can always transform a correct SEMIPAL book into a correct PAL book by the following device. Find a symbol, Q say, that does not yet occur as an identifier. Write the line

0	Q	:=	PN	<u>type</u>

and let it be followed by the SEMIPAL book where we add Q as category of each line.)

4.3 The rules of SEMIPAL are given in this and the next section. The reader may take the 13 lines of sec.4.1 as an example, by just cancelling the category column.

(i) As the first line of the book any one of the lines

0	...	:=	PN
0	...	:=	—

is acceptable. (Here "..." stands for an arbitrary identifier.)

(ii) We can add an (n+1)-st line to a correct SEMIPAL book Λ of n lines by writing

$$u \quad \ldots \quad := \quad \Sigma ,$$

where u is either 0 or one of the previous block openers, and Σ is either —, or PN, or an expression valid at u, a notion to be defined presently.

4.4 The notion <u>expression valid at u</u> is relative to the given correct book Λ. We define it by recursion.

(1) If b is a block opener, either equal to u or contained in the indicator string of u, then b is an expression valid at u. Example: At y the expressions n,y are valid.

(2) If b is the identifier of a line of Λ, but not a block opener, and if the indicator of that line is either 0 or u or contained in the indicator string of u, then b is an expression valid at u. Example: At y the expressions nat, real, d, e, f, g are valid.

(3) Let b the identifier part of one of the lines of Λ, and assume that b is not a block opener. Let n be the length of the indicator string of b. Let k be a second integer, $0 < k \leq n$. We assume that $\Sigma_1, \ldots, \Sigma_k$ are expressions valid at u. If $n > k$ we have the extra assumption that the (n-k)-th entry of the indicator string of b is an expression valid at u (that is, it is equal to u or contained in the indicator string of u). Then

$$b(\Sigma_1, \ldots, \Sigma_k)$$

is an expression valid at u.

4.5 In the SEMIPAL book that is obtained from the example of sec. 4.1 (by omitting the category column) we give a few examples of expressions valid at y:

n; y; f; prod(d,f); e(d); power(n,f); power(f); d(y); d(n,y); e(prod(e,e)).

4.6 As a preparation to discussion of normal forms, we define the <u>completion</u> of an expression valid at u Let Σ be an expression valid at u; its completion Σ' will also be valid at u.

(i) If Σ consists of a single block opener, then $\Sigma' = \Sigma$.

(ii) Let $\Sigma = b(\Sigma_1,\ldots,\Sigma_k)$ (see the end of sec. 4.4) and let $u_1,\ldots,u_n$ be the indicator string of b. Then

$$\Sigma' = b(u_1,\ldots,u_{n-k}, \Sigma_1,\ldots,\Sigma_k).$$

If $k = 0$, $n - k \neq 0$ this has to be read as $b(u_1,\ldots,u_{n-k})$, if $k \neq 0$, $n - k = 0$ as $b(\Sigma_1,\ldots,\Sigma_k)$, if $k = n - k = 0$ it has to be read as just b.

4.7 An expression is said to have <u>normal form</u> (in the sense of SEMIPAL) if it contains no compound notions (see sec. 3.2).

Let Σ be an expression valid at u. We shall define, again recursively, a reduction to normal form Σ^*. We first complete the expression Σ to Σ' (4.6).

If Σ' is a single identifier, but not a compound notion, then we take $\Sigma^* = \Sigma'$.

If Σ' is a single identifier and if that identifier is a compound notion, we define Σ^* to be the normal form of Ω, where Ω is the definition part of the line whose identifier part is Σ'.

If $\Sigma' = b(\Sigma_1,\ldots,\Sigma_n)$ with $n > 0$, and if b is a primitive notion, then we take

$$\Sigma^* = b(\Sigma_1^*,\ldots,\Sigma_n^*),$$

where Σ_i^* is the normal form of Σ_i $(i = 1,\ldots,k)$.

If $\Sigma' = b(\Sigma_1,\ldots,\Sigma_n)$ with $n > 0$, and if b is a compound notion, with indicator string $u_1,\ldots,u_n$, then we obtain Σ^* as follows. Let Ω^* be the normal form of the definition part of the line whose identifier is b. In Ω^* replace every occurrence of u_i by Σ_i^* (the normal form of Σ_i). This gives Σ^*.

Warning: the substitution of the Σ_i^* for u_i is only carried out for explicit occurrences of u_i in Ω^*, and not for new u_i's that arise <u>after</u> substitution (the Σ_i^*'s themselves may contain u_j's).

As an example we give the normal form of the expression e(d) of line 13 in the example of sec. 4.1:

prod(power(n,power(n,y)),power(n,y)).

4.8 Two expressions Σ_1, Σ_2 both valid at u are called <u>definitionally equivalent</u> if they have the same normal form. If we want to show definitional equivalence it is not always necessary to compute these normal forms; it will often suffice if we can transform both forms into a single form by partial reduction.

If we replace an expression in a correct SEMIPAL book by a definitionally equivalent one, we get a new correct SEMIPAL book. The normal forms of corresponding expressions in both books will be the same.

4.9 We shall describe the notion of a correct PAL book in two stages. We start with a book Λ written according to the preliminary description of sec. 3. That is, the definition part of a line is —, or PN, or an expression; the category part is <u>type</u> or an expression; the indicator part is 0 or a previous block opener. By a certain duplication operation to be described presently, we get something which we shall require to be a correct SEMIPAL book Λ'. Finally, we shall require certain conditions regarding the categories.

The duplication means the following thing. We replace every line

$$u \quad a \quad := \quad \Omega \quad \Sigma$$

(where u may be 0 or a block opener, Ω may be an expression or — or PN, Σ is an expression or <u>type</u>) by two lines

$$u \quad a^{+} \quad := \Sigma$$
$$u \quad a \quad := \Omega \, ,$$

unless Σ is <u>type</u>, in which case we write the single line

$$u \quad a \quad := \Omega \, .$$

We of course assume that for every identifier we can create an entirely new identifier by adding the plus sign.

As an example we deal with the first 5 lines of the book of sec. 4.1:

0	nat	:=	PN
0	real	:=	PN
0	a^+	:=	real
0	a	:=	—
a	b^+	:=	real
a	b	:=	—
b	$prod^+$	:=	real
b	prod	:=	PN

4.10 We define the notion "correct PAL book" by induction. The definition will be such that if Λ is correct, then Λ' is a correct SEMIPAL book.

A one-line book is correct if and only if that line has one of the following two forms:

0	...	:=	PN	<u>type</u>,
0	...	:=	—	<u>type</u>.

Now assume that a book Λ consisting of n lines is a correct PAL book. We shall state the conditions for any line to be added.

(i) The indicator u is a block opener of Λ.

(ii) The definition part is either —, or PN, or an acceptable expression at u (see sec. 4.11 for this).

(iii) The category part is either <u>type</u>, or an acceptable expression at u with category <u>type</u>. In the case where the definition part is an expression, (see(ii)), we require that the category part is definitionally equivalent (in the sense of the SEMIPAL book Λ') to the category of that expression.

4.11 Let u be one of the block openers of the SEMIPAL book Λ' obtained by duplication of Λ. We will define a collection of expressions that we call <u>acceptable</u> at u; to each one of these expressions we will attach what we will call a <u>category</u>. The latter is either an expression or the symbol <u>type</u>. The expressions to be considered will only contain identifiers of Λ, and no identifiers with plus signs attached to them. The acceptable expressions will be automatically valid at u in the sense of sec. 4.4.

The description of "acceptable" closely resembles the one of "valid".

(1) Let b be one of the following:

a block opener whose indicator string is contained in the indicator string of u;

the identifier of a line of Λ (but not a block opener) whose indicator is either 0, or u, or contained in the indicator string of u (cf. (1) and (2) in sec. 4.4).

Then b is an acceptable expression at u, and its category is the category part of the line whose identifier is b.

(2) Let b be the identifier part of one of the lines of Λ, and assume assume that b is not a block opener. Let n be the length of the indicator string of b. Let k be a second integer, $0 \leq k \leq n$. We assume that the expressions $\Sigma_1,\ldots,\Sigma_k$ are acceptable at u, with categories $\Omega_1,\ldots,\Omega_k$. If $n > k$ we have the extra condition that the (n-k)-th entry of the indicator string of b is either equal to u or contained in the indicator string of u. Let $v_1,\ldots,v_k$ be the last k entries in the indicator string of b. We require, for $i = 1,\ldots,k$, that

$$v_i^+ (\Sigma_1,\ldots,\Sigma_{i-1}) \qquad (1)$$

is definitionally equivalent (in the sense of Λ') to Ω_i. (If $i = 1$ we have to read (1) as v_1^+. If any of the v_i^+ does not occur in Λ', we have to read (1) as <u>type</u>, and the condition is just that Ω_i = <u>type</u>.)

Under these conditions we proclaim $b(\Sigma_1,\ldots,\Sigma_k)$ to be acceptable at u, and we give it the category $b^+(\Sigma_1,\ldots,\Sigma_k)$. If b^+ does not occur in Λ', the new expression $b(\Sigma_1,\ldots,\Sigma_k)$ is given the category <u>type</u>.

One minor modification should be made: we promised that the category would not be an expression containing identifiers with plus signs. Therefore we replace $b^+(\Sigma_1,\ldots,\Sigma_k)$ by the result of an application of a substitution such as described at the end of sec. 4.7.

<u>How to use PAL for mathematical reasoning.</u>

In section 4 we explained how to express things by means of PAL. Seemingly, expressing things covers only a small part of mathematics, for usually we are interested in proving statements. Mathematics has the same block structure as we have in PAL, but there are two ways to open a block.

One is by introducing a variable that will have a meaning throughout the block, the other one is by making an assumption that is valid throughout the block. We shall be able to deal with the second case as efficiently as with the first one, if we represent statements by categories. Saying we have a thing in such a category means asserting the statement. This can be done in three ways: by means of —, or PN, or an expression. These three correspond to assertion by assumption, by axiom, by proof, respectively.

5.2 As an example we shall deal with equality in an arbitrary category. The following piece of text introduces equality as a primitive notion, and states the three usual axioms.

0	ξ	:=	—	<u>type</u>	1
ξ	x	:=	—	ξ	2
x	y	:=	—	ξ	3
y	is	:=	PN	<u>type</u>	4
x	reflex	:=	PN	is(x,x)	5
y	asp 1	:=	—	is(x,y)	6
asp 1	symm	:=	PN	is(y,x)	7
asp 1	z	:=	—	ξ	8
z	asp 2	:=	—	is(y,z)	9
asp 2	trans	:=	PN	is(x,z)	10

This book is <u>not</u> a nested one since line 5 does not belong to the block opened by y. Even so, the vertical bars, with an interruption at line 5, can be helpful.

We now show how this piece of text can be used in later parts of the book. Assume we have the following lines (in some order) in the book:

0	η	:=	....	<u>type</u>
0	a	:=	....	η
0	b	:=	....	η
0	known	:=	....	is(η,a,b)

We wish to derive a line:

0 result := .... is(η,b,a).

We have to find a definition part for this line. What we want is to apply line 7. The indicator string is (ξ,x,y,asp1). In ordinary mathematical terms, we have to furnish a value for ξ, a value for x, a value for y, and a proof for the statement obtained from "x = y" by these substitutions. A proof for the statement means, in our present convention, something of the category is(η,a,b). Indeed we have something, viz. "known". The reader can easily verify that

0 result := symm(η,a,b,known) is(η,b,a)

is an acceptable line.

The above application was given entirely in context 0, but it can be done in any block that contains η,a,b and known.

5.3 We are, of course, inclined to see the categories as classes, and things having that category as elements of those classes. If we want to maintain that picture, we have to say that the category "is(ξ,x,y)" consists of all proofs for x = y. In this picture the usual phrase "assume x=y" is replaced by "let p be a proof for x=y". Another aspect is that we have to imagine the category "is(ξ,x,y)" to be empty if the statement x=y is false. The latter remark points at a difference between these assertion categories and the "ordinary" categories like "nat" and "real" in sec. 4. In the spirit of the example of sec. 4 it is vital to know what the expressions are, and it seems pretty useless to deal with empty categories. With the assertion categories it is different. The interesting question is whether we can find something in such a category, it doesn't matter what.

5.4 A modern mathematician has the feeling that asserting is something we do with a proposition. The author thinks that this is not the historic point of view. The primitive mathematical mind asserts the things it can, and is unable to discuss things it cannot assert. To put it in a nicer way, it has a kind of contructivist point of view. It requires a crooked way of thinking to build expressions that can be doubted, i.e. to build things that might or might not be asserted. A possible way to do this in PAL is to talk about the category "bool" consisting of all propositions, and to attach to each proposition an assertion category. We start the book like this:

0	bool	:=	PN	<u>type</u>
0	| b	:=	—	bool
b	| TRUE	:=	PN	<u>type</u>

The standard interpretation is simple. If we write in a certain context

... := .... TRUE(c),

where c is (in that context) a proposition, then the interpretation in every-day mathematical language is that we are asserting c.

5.5 In PAL we are able to write axioms and prove theorems about propositions (e.g. tautologies). In later parts of the book we will be able to use these axioms and theorems (just like the derivation of "result" in sec. 5.2). This means that in a PAL book we are able to derive inference rules that can be applied later in that same book.

As a very primitive example we shall write the following in PAL. After introducing bool and TRUE we introduce the conjunction of two propositions. We present some axioms concerning that conjunction, and we show that from x ∧ y we can derive y ∧ x. Finally we show how in a later piece of text the result can be used as an inference rule.

0	bool	:=	PN	<u>type</u>
0	| b	:=	—	bool
b	| TRUE	:=	PN	<u>type</u>
0	| x	:=	—	bool
x	| | y	:=	—	bool
y	| | and	:=	PN	bool
y	| | | asp 1	:=	—	TRUE(x)
asp 1	| | | | asp 2	:=	—	TRUE(y)
asp 2	| | | | ax 1	:=	PN	TRUE(and)
y	| | | asp 3	:=	—	TRUE(and)
asp 3	| | | ax 2	:=	PN	TRUE(x)
asp 3	| | | ax 3	:=	PN	TRUE(y)
asp 3	| | | theorem	:=	ax 1(y,x ax 3, ax 2)	TRUE(and(y,x))

0	u	:=	....	bool
0	v	:=	....	bool
0	known	:=	....	TRUE(and(u,v)
0	derived	:=	theorem(u,v,known)	TRUE(and(v,u))

5.6 The reader will have observed from the above examples that we do not need to subdivide our text into parts like "theorem", "proof", "definition", "axiom". Every line is a result that can be used whenever we wish. It may require a large number of lines to translate the proof of a theorem into PAL. (Needless to say, we can always try to reduce the number of lines, but that makes the lines more complicated and hard to read.) Some of the lines represent definitions of notions introduced only for the sake of the proof. Other lines represent sub-results, usually called lemmas. The usual idea about theorems and proofs is, at least formally, that we are not allowed to refer to results obtained inside a proof. In PAL (and in AUTOMATH), however, we are free to use every line everywhere. We never announce a theorem before the proof starts, the result cannot be stated before it has been derived.

6. Extending PAL to AUTOMATH.

6.1 It was shown in sec. 4 how we can deal with functional relationship in PAL. Once a function has been defined (either by PN or by definition in terms of previous notions) it can be applied. That is, a function f is introduced by saying what the value of f(x) is for every x of a certain category. And if we have, at a later stage, an expression Σ having that same category, it will be possible to talk about $f(\Sigma)$. A thing that we can not write in PAL, however, is "let f be any function, mapping category Σ_1 into category Σ_2". If we wish to deal with such mappings the way it is done in mathematics, we want several things:

(i) We need the facility of building the category of the mappings of Σ_1 into Σ_2.

(ii) If f is an element of that mapping category, and if x is something having category Σ_1, then we have to be able to form the image of x under f.

(iii) If a mapping of Σ_1 into Σ_2 is explicitly given in the PAL way then we have to be able to recognize that mapping as a member f of the mapping category.

(iv) If we apply (ii) to the f obtained in (iii), we can (making x a block opener) obtain a function given in the PAL way. This function should be equivalent to the one we started from in (iii).

6.2 Let us consider (iii) more closely. The "PAL way" of giving a function is the following one: We have somewhere in the book

u	x	:=	—	Σ_1	1
x	v	:=	Λ	Σ_2	2

where Λ is an expression possibly depending on x. (That is, its normal form may contain x.) But it is only fair to remark that Σ_2 may also depend on x; Σ_1, on the other hand, can <u>not</u> contain x. Let us assume that neither Σ_1 nor Σ_2 is the symbol <u>type</u>.

The mapping described here attaches to every x of type Σ_1 a value depending on x, which value has category also depending on x. We shall use the notation

$$[x,\Sigma_1]\Sigma_2$$

for the category of this mapping, and

$$[x,\Sigma_1]\Lambda$$

for the mapping itself. There is an objection against using the old identifier x for this new purpose, and therefore we replace it by a new identifier t. This t will never occur as identifier part of a line. It is called a <u>bound variable</u>, and we may assume that it will be used here, but never again.

We shall write $\Omega_x(\Sigma)\Lambda$ for the result of substitution of Σ for x in the expression Λ. (It should be remarked that Λ may contain x implicitly. In order to make such implicit occurrences explicit, we have to transform Λ by application of definitions up to a point where further implicit occurrence is impossible, since we left the block where x is valid. This substitution operation seems to be harder than the corresponding one in PAL where we could express ourselves in terms of normal forms. In practice, however, it does not make much of a difference; normal forms in PAL are only of theoretical interest.)

We can now phrase the rule of functional abstraction: In AUTOMATH we have the right to deduce from lines 1 and 2 the acceptability of the line

u	...	:=	$[t,\Sigma_1]\Omega_x(t)\Lambda$	$[t,\Sigma_1]\Omega_x(t)\Sigma_2$	3

Accordingly we have the right to consider $[t,\Sigma_1]\Omega_x(t)\Sigma_2$ as a category. So if we have (if Σ_1 and Σ_2 are expressions)

u	x	:=	—	Σ_1	4
x	w	:=	Σ_2	type	5

we have the right to add

u	...	:=	$[t,\Sigma_1]\Omega_x(t)\Sigma_2$	type	6

This makes it possible to open a new block with

u	f	:=	—	$[t,\Sigma_1]\Omega_x(t)\Sigma_2$,	7

that is, we can start an argument with: let f be any mapping of the described kind. We also have the possibility to write line 7 with PN instead of —.

6.4 Now returning to point (ii) sec. 6.1, we introduce the following rule. If we have a line

u	...	:=	Γ	$[t,\Sigma_1]\Omega_x(t)\Sigma_2$	8

and also a line

u	...	:=	Δ	Σ_1

then we take the liberty to write

u	...	:=	$\{\Delta\}\Gamma$	$\Omega_x(\Delta)\Sigma_2$.

The interpretation is that $\{\Delta\}\Gamma$ is the result of the substitution of Δ into Γ. We write this instead of $\Gamma(\Delta)$ since, in the case that Γ is a single identifier, the latter notation already had an entirely different meaning in PAL: it was used to change context. That is, $\Gamma(\Delta)$ is the mapping we obtain from Γ if we substitute Δ for u, and it is even questionable whether this is possible, since u need not be of category Σ_1.

6.5 In connection with this notation $\{\ \}$ we take the liberty to extend the notion of definitional equality by the following pair of rules:

(i) If $\Sigma_1, \Sigma_2, \Sigma_3$ are expressions, where Σ_2 contains the bound variable t, but Σ_1 and Σ_3 do not, then we postulate the definitional equality of

$$\{\Sigma_3\}\ [t, \Sigma_1]\ \Sigma_2 \qquad \text{and} \qquad \Omega_t(\Sigma_3)\,\Sigma_2.$$

That is, it does not make a difference whether substitution is carried out <u>before</u> or <u>after</u> functional abstraction.

(ii) If Σ_1 and Σ_2 are expressions that do not contain the bound variable x, then we postulate the definitional equality of

$$[x, \Sigma_1]\ \{x\}\Sigma_2 \qquad \text{and} \qquad \Sigma_2.$$

The above rules (i) and (ii) explain why we prefer to write $\{x\}f$ instead of $f\{x\}$. By way of these rules, $\{x\}\, f$ is in agreement with the convention $[t,\Sigma_1]\Sigma_2$ for functional abstraction, and the latter is in agreement with the general mathematical habit to write quantifiers like

$$\forall_{x \in S},\quad \cup_{x \in S},\quad \Pi_{n=1}^{\infty}$$

on the left of the formulas they act on.

6.6 The description of AUTOMATH in the preceding sections was not as complete as the description of SEMIPAL and PAL in sec. 4. For a complete and more formal definition of AUTOMATH we refer to the report mentioned in sec. 1.2.

7. How to use AUTOMATH for mathematical reasoning.

7.1 If we write elementary mathematical reasoning in PAL as described in section 5, one of the first things we can not do is to derive an implication . There are two things we wish to do with implication, and only one of the two can be done in PAL.

First assume we have introduced implication as a primitive notion, then it is easy to write "modus ponens" as an inference rule:

0	bool	:=	PN	type	1
0	b	:=	—	bool	2
b	TRUE	:=	PN	type	3
b	c	:=	—	bool	4
c	impl	:=	PN	bool	5
c	asp 1	:=	—	TRUE(b)	6
asp 1	asp 2	:=	—	TRUE(impl)	7
asp 2	modpon	:=	PN	TRUE(c)	8

By means of this piece of text we are able to use the inference rule

$$\frac{A,\ A \Rightarrow B}{B}$$

in all possible situations.

The second thing we want to do is this. If we have

0	p	:=	....	bool	9
0	q	:=	....	bool	10
0	asp 3	:=	—	TRUE(p)	11
asp 3	then	:=	....	TRUE(q)	12

(it might have been given in any other context instead of 0) then we want to construct s mething in TRUE(impl(p,q)). This cannot be done by means of the rules of PAL.

The problem can be solved in AUTOMATH, however. We first say that if we have a mapping from TRUE(b) into TRUE(c), then impl(b,c) is true:

```
c          |||asp 4     :=    —                [x,TRUE(b)] TRUE(c)   13
asp 4      |||axiom     :=    PN               TRUE(impl)            14
```

Using the axiom, and functional abstraction, we can derive from lines 11, 12

```
0          first        :=  [y,TRUE(p)] then(y)   [y,TRUE(p)] TRUE(q)   15
0          second       :=  axiom(p,q,first)      TRUE(impl(p,q))       16
```

That is, we have derived an assertion of impl(p,q). So we have the inference rule

$$\frac{\frac{A}{B}}{A \Rightarrow B}$$

available in all possible cases.

If we wish, we can write the application of this inference rule in one line instead of two, viz.

```
0        ...     :=   axiom(p,q[y,TRUE(p)]then(y))   TRUE(impl(p,q))   17
```

7.2 As a second example we introduce the all-quantifier for a predicate P on an arbitrary type ξ.

```
0       bool       :=   PN    type                 1
0       |b         :=   —     bool                 2
b       |TRUE      :=   PN    type                 3
0       |ξ         :=   —     type                 4
ξ       ||P        :=   —     [u,ξ]bool            5
P       ||all      :=   PN    bool                 6
P       |||x       :=   —     ξ                    7
x       ||||asp 5  :=   —     TRUE(all)            8
x       ||||ax 1   :=   PN    TRUE({x}P)           9
P       |||asp 6   :=   —     [v,ξ] TRUE({v} P)   10
asp 6   ||||ax 2   :=   PN    TRUE(all)           11
```

Note the close resemblance between the text of sec. 7.1 and this one. Actually we are able to define "impl" in terms of "all": We can write instead of line 5 of sec. 7.1

c   impl   :=   all(TRUE(b), [t,TRUE(b)] c)   bool

If we do this after having accepted the text of 7.2, then we can replace the PN's in line 8 and line 14 of sec. 7.1 by proofs. The reader may check that the PN in line 8 (sec. 7.1) can be replaced by

ax1(TRUE(b),[s,TRUE(b)] c, asp 1, asp 2),

and the one in line 14 (sec. 7.1) by

ax 2(TRUE(b), [s,TRUE(b)]c, asp 4).

7.3 Next we discuss the existence quantifier. There are various different approaches to this. The simplest one, and therefore the easiest one for application, is connected with the Hilbert operator. It says, if for any given category there exists an object for which a given property holds, then we have a way of selecting such an object as if we were in possession of a standard algorithm that selects for us.

We can write this as follows. We start again with the introduction of bool and TRUE, then we take an arbitrary category ξ and an arbitrary predicate on that category, and we introduce existence as a primitive notion. It says that "existence" is true if and only if we have something in that category ξ.

0	bool	:=	PN	<u>type</u>	1
0	| b	:=	—	bool	2
b	| TRUE	:=	PN	<u>type</u>	3
0	| ξ	:=	—	<u>type</u>	4
ξ	| | P	:=	—	[u,ξ] bool	5
P	| | exists	:=	PN	bool	6
P	| | | v	:=	—	ξ	7
v	| | | | asp 1	:=	—	TRUE({v}P)	8
asp 1	| | | | axiom 1	:=	PN	TRUE(exists)	9
P	| | | asp 2	:=	—	TRUE(exists)	10
asp 2	| | | Hilbert	:=	PN	ξ	11
asp 2	| | | axiom 3	:=	PN	TRUE({Hilbert}P)	12

In combination with other axioms this way of defining existence easily leads to non-contructive things, e.g. the axiom of choice.

A different way of introducing existence is to say that it is not true that the negation of the predicate holds for all objects in the given category. This of course requires a definition of negation, which can be done in several ways. We shall not discuss it here.

The difficulties about existence arise already at a lower level, viz. with the notion of non-emptiness of a category. In that case the following may be a useful substitute for the kind of non-emptiness related to the Hilbert operator:

0	ξ	:=	—	<u>type</u>
ξ	NEPTY	:=	[c,bool][u, [x, ξ]TRUE(c)]TRUE(c)	<u>type</u>

So if we have something in NEPTY, and if c is any proposition, and if we can prove that whenever we have an x in ξ then c is true, then we have proved c. So if we have something in NEPTY, we have a kind of inference rule: If we want to prove a proposition c then we may act as if we know an x with category ξ.

7.4 There is no objection against higher order predicate calculus in AUTOMATH. For example, we can talk about the category R of all predicates on the category of natural numbers say, about the category S of all predicates on R, etc.:

0	nat	:=	....	<u>type</u>
0	R	:=	[n,nat]bool	<u>type</u>
0	S	:=	[r,R]bool	<u>type</u>

7.5 Every language has its advantages and disadvantages. The disadvantages of AUTOMATH are obvious: it is tedious to have to write in full detail, carefree identification of things in different categories is forbidden (see sec.2.2), and embedding of types into other types is not an automatic facility. In order to compensate for these disadvantages, the user should try to exploit the advantages the language has. One advantage is that we do not have to announce theorems and lemmas in a formal way, and therefore repetition of arguments is much easier suppressed than in ordinary mathematics. And, of course, we can invent all sorts of tricks, We present just one such trick here.

Consider an axiom like the line TRUE in sec. 5.4. Once we have written it this way, we cannot get rid of it: if we want to do mathematics without it, we have to write a new book. There is a way, however, to introduce the axiom in such a way that, so to speak, it is only available to those who have authority to use it. We introduce a new primitive notion AUTH (for authority) and then state the axiom for those users who have something in AUTH:

0	bool	:=	PN	<u>type</u>
0	AUTH	:=	PN	<u>type</u>
0	| a	:=	—	AUTH
a	| | b	:=	—	bool
b	| | TRUE*	:=	PN	<u>type</u>

If later we have c in AUTH and d in bool, we can use TRUE*(c,d). If c in AUTH is valid in a large part of the book, we can get rid of the awkward obligation to mention our authority, by defining (in a context where c is available)

| e	:=	—	bool
| TRUE	:=	TRUE*(c,e)	<u>type</u>

and now we can write TRUE(f) for any proposition f.

8. Unsolved problems about AUTOMATH.

8.1 It is very probable (but not yet proved) that the following is true. If the lines

u	...	:=	Σ_1	Λ_1
u	...	:=	Σ_2	Λ_2

occur in a book, if Σ_1 and Σ_2 are definitionally equivalent, then Λ_1 and Λ_2 are definitionally equivalent. We only say roughly what definitional equivalence is: Two expressions are definitionally equivalent if one of them can be transformed into the other by replacing an identifier in one of the expressions by the expression that defines it, and also by application of one of the operations of the lambda calculus. They are also called definitionally equivalent if they can be connected by a chain of pairwise definitionally equivalent expressions.

We do not express the notion by means of normal forms, as in 4.7, since we are not yet sure about normal forms.

8.2 Probably every expression occurring in a AUTOMATH book is definitionally equivalent to an expression that does not contain any } followed by a [. This means an expression

$$[\beta_1, \Sigma_1] \ldots [\beta_k, \Sigma_k] \{\Gamma_1\} \ldots \{\Gamma_h\} \beta (\Theta_1, \ldots, \Theta_m)$$

(possibly k = 0, h = 0, or m = 0), where the Greek capitals again represent expressions of that form, the $\beta_1, \ldots, \beta_k$ are bound variables, and β is either a block opener or the identifier part of a line with PN.

9. Processors for AUTOMATH.

9.1 A processor is a computer program that enables a computer to check line by line whether any given input represents a correct AUTOMATH book.

One of the things the computer gets to do is to check whether two expressions are definitionally equivalent. Even if the conjectures of sec. 8 are true, it can be very impractical to use normal forms for checking that equivalence. It is already impractical in PAL, where there is no difficulty with the normal forms (see sec. 4.7).

A good processor should have a good strategy for checking equivalence. In cases where the general strategy is failing, it may pay to assist the computer by giving hints as to what to do first.

It is to be expected that very few hints will be needed in general. That is, at least as long as we do not try to condense a larger number of lines into a single one. Such a condensation is quite often possible, it saves identifiers, but makes things harder to write and harder to check. (An additional disadvantage of condensed writing is the repetition of expressions which might have been abbreviated by means of extra lines.

Another aspect of the same thing is giving an argument twice where a lemma might have been more efficient.)

9.3 There are several attractive possibilities for man-machine interaction if a terminal is available for direct communication in conversational mode. (The AUTOMATH processor in operation in 1968 at the Technological University, Eindhoven, did not yet provide such facilities.) For lines the machine rejects, it can produce diagnostics by means of which the operator can carry out corrections or add hints. It will be very practical for the operator to suppress the category of a line (unless the definition is — or PN), and to ask the machine what category it finds. If this does not coincide with the one the operator has in mind, the operator can ask the machine to check definitional equivalence of the two expressions.

10. Possibilities for superimposed languages.

10.1 For practical purposes it will be attractive to make languages which bear the same relation to AUTOMATH as a programming language has to some particular machine language. We shall call such languages superimposed on AUTOMATH. They require a compiler for translation into AUTOMATH.

10.2 A very simple thing a superimposed language might do is admitting repetition of names (such as the repeated use of the letter x for many different purposes in the book). The compiler has to rename everything in order to meet the requirement that in AUTOMATH the identifier parts of the lines are distinct.

10.3 In more complicated cases the superimposed language will require a fixed correct AUTOMATH book as a basis. If we have written a book in the superimposed language, then the compiler starts from the basis, and next it translates the given book into AUTOMATH lines which are subsequently added to the basis, and checked by the AUTOMATH processor.

10.4 In a superimposed language standard mathematical notation might be used more freely. For example, in the superimposed language one might write p := a + b + c. The compiler sees that a,b,c were previously introduced as reals, it sees that no change of context has been mentioned, it knows that "real" and "plus" are identifiers in the basis. It writes

p := plus(plus(a,b),c)   real

and it keeps the context indicator of the previous line.

10.5 A superimposed language might be very different from AUTOMATH in its approach to things like propositions, assertions, predicates. The user of the superimposed language need not even notice that AUTOMATH has a slightly unconventional approach to these things.

10.6 It is not strictly necessary that the text presented in a superimposed language is entirely unambiguous and free of gaps. Just as the human mathematician has been trained to guess what the sentences in his textbook mean exactly, the compiler can be trained to guess the meaning of what is said in the superimposed language. It cannot be expected to do very much in this direction, but whatever it can do, will be very helpful. Writing absolutely meticuously is very much harder than writing almost meticuously, and it will be a great gain if a machine can bridge the gap between the two.

11. Automatic theorem proving.

11.1 AUTOMATH is not intended for automatic theorem proving. Theorem proving is a difficult and time-consuming thing for a machine. Therefore it is almost imperative to devise a special representation of mathematical thinking for any special kind of problem. Using a general purpose language like AUTOMATH would be like using a contraption that is able to catch flies as well as elephants and submarines.

11.2 There is a case for automatic proof writing in AUTOMATH if we have to produce a tedious long proof along lines that can be precisely described beforehand. Let us take an example. Assume that P is a proposition on magic squares, and that we want to prove a theorem saying that there is no 8×8 magic square that has property P. We can write a computer program for this and run it on a computer. The computer says that none exist. Now quite apart from the question whether the computer is right, we have to admit that a formal mathematical proof has not been produced. Even if we had a complete mathematical theory about the machine, the machine language, the programming language, our proof would depend on intuitive feelings that the program gives us what we want, and it would definitely depend on a particular piece of hardware.

For those who are willing to take AUTOMATH, at least temporarily, as their only final conscience of mathematical rigour, there is a way out. We can rewrite the magic square program in such a way that the search is stepwise accompanied by the production of AUTOMATH lines that give account of a detailed mathematical reasoning, ending with the conclusion that there is no 8×8 magic square with property P. This way we get a complete proof that can be checked by any mathematician. If we leave the checking to a computer, again we get into the question of whether the processor and the computer do what we expect them to do, but that is an entirely different matter.

. Extensions of AUTOMATH.

.1 If we feel we should have a more powerful language than AUTOMATH, this can have two reasons.

.2 One reason is that we feel that the language is clumsy, and that we want to make it more handy, without changing the scope of what we can say. For some purposes this might be possible by extension of the language, i.e. by adding new grammar rules without cancelling the old ones. It is hardly necessary to consider such extensions for the present purpose, since it can be expected that the same goal can be reached by means of superimposed languages. We might think about facilities for easy identification of two things of different categories (see sec. 2.2), embedding of one category into another, etc. If such matters can be handled satisfactorily, they can be handled by a superimposed language. The only reasons for doing it without such a language may be computer time and memory space.

12.3 A different reason for extension can be that we feel that AUTOMATH is not strong enough, just as we extended PAL to AUTOMATH since PAL was not strong enough for modern mathematics.

One might suspect that no single language will ever be entirely satisfactory. It is an old mathematical habit to mix language and metalanguage: we write a text in a language; we discover facts about that text; we use these facts in the subsequent text. This of course means an extension of the language. We mention an example, though not a very important one. Let q be any identifier in an AUTOMATH book, and let p be a block opener. If it happens that q does not implicitly depend on p, this is an observation <u>about</u> the book, and there seems to be no way to write it as an assertion <u>in</u> the book. It will be an extension of the language if we design some way to write this independence, a way to derive it from the book, and a way to use that written information if we need it. This kind of thing is done in ordinary mathematical language, but in AUTOMATH it is not necessary. If q does not depend on p, then we are able to define r := q in a context where p is not valid, and then need not bother about p any more.

12.4 There is a class of extensions of AUTOMATH that is very easy to describe: We start the book with a number of lines some of which have not been written according to the rules; we want to write the rest of the lines in the book according to the rules. We give an example that does not belong to AUTOMATH, but to the language we get from AUTOMATH if PN's are forbidden: Then we can write all axioms in the basis as theorems without proofs, and talk PN-free language ever after.

One might even think of an infinitely long basis. For example, one might like to have all the natural numbers as <u>a priori</u> given, and devote a line or two to each one of them.

12.5 In AUTOMATH we have the right to indulge in functional abstraction with respect to every type. In private discussions Prof. Dana Scott said he did not like the idea of introducing "bool" as such a type, at least not in intuitionism. It is very easy to extend AUTOMATH by introducing a symbol <u>type*</u>, and saying that if Σ has category <u>type*</u>, then we do not have the right of functional abstraction with respect to Σ . It seems fair to admit the category $\Sigma_3 := [x,\Sigma_1]\Sigma_2$ if Σ_1 has category <u>type</u> and Σ_2 has

has category _type*_, and to say that Σ_3 has category _type*_. If we do all this, we can introduce "bool" as something of category _type*_, and "nat" (the natural numbers) as something of category _type_.

12.6 In AUTOMATH we did not allow functional abstraction with respect to _type_ itself. For example, if we have

0	ξ	:=	—	_type_
ξ	b	:=	PN	bool

then we can _not_ write

0	...	:=	[t,_type_] b(t)	[t,_type_] bool.

It is difficult to see what happens if we admit this.

12.7 A possibility that seems less dangerous than the one of 12.6 is the following one: if we have

0	ξ	:=	....	_type_
0	a	:=	—	ξ
a	b	:=	PN	_type_

then we allow to write

0	...	:=	[t,ξ]b(t)	[t,ξ]_type_

This gives more information about [t,ξ]b(t) than just saying that it has category _type_, but on the other hand it puts an end to uniqueness of category.

Moreover, we permit lines such as

0	a	:=	—	[t,ξ] _type_

in order to introduce an arbitrary way of attaching a type to each t in ξ.

Once we have opened these possibilities, it will be pretty obvious what the further operational rules have to be.

We mention a single case where this extension of our language is needed. In connection with recursive definitions, we might wish to say: let $P_1, P_2, \ldots$ be an infinite sequence of categories. This can be done by means of a block opener with category [n,nat] _type_.

PROOF THEORY AND THE ACCURACY OF COMPUTATIONS *)

Erwin Engeler

Imagine a program π_1 for the solution of, say, a system of linear equations. The matematical work that goes into obtaining π_1 makes use of a body of knowledge about the field $\underset{\sim}{R}$ of real numbers. And indeed, if the computations according to π_1 were performed by an ideal computer working directly with the reals, i.e. with infinite accuracy, then the computed values would actually be solutions. In reality, however, the program π_1 is executed on a less than ideal computer which works with some sort of trunctuated reals. Thus it may happen that the computed values are in fact not solutions at all.

The problem that is posed by this situation is to characterize those programs whose meaning is preserved under passage from the ideal to the actual computer. The remaining remarks in this introduction are designed to make the formulation of this problem more precise.

If π_1 is a program for the solution of some mathematical question we can in general find a program π_2 which checks whether the values computed by π_1 actually form a solution. In the case of systems of linear equations such a checking program can be arranged so that the composition of programs

$$\rightarrow \pi_1 \rightarrow \pi_2 \rightarrow$$

is a program that terminates on an input iff π_1 computes a solution. In other case, for example if π_1 computes a sequence of approximations to $\sqrt{2}$ by nested intervals, the checking is arranged as

$$\to \pi_1 \to \pi_2 \to$$

and does <u>not</u> terminate iff the sequence of computed values converges to $\sqrt{2}$.

All that we can reasonably expect of an actual computer is that it obtains results that are faithful within the limits of its discerning powers. Thus we are lead to reformulate our question, provisionally, into: what are the programs whose termination (or non-termination) is preserved under passage from the ideal to the actual computer?

In order to approach this problem we need some information about the relation between the actual and the ideal computer. For the present we make the, idealizing, assumption that the system of "trunctuated reals" on which the actual computer operates constitutes a homomorphic image of the field $\underline{R}$ of real numbers. The problem before us is therefore: What are the programs whose termination or non-termination is preserved under homomorphisms?

A convenient framework in which to treat questions of this nature is the general theory of machines and its interconnections with infinitary logic established in [2] and developed in [3]. Namely, termination and non-termination of programs π can be expressed by a logical formula $\varphi(\pi)$ which can be effectively obtained from π. The language to which $\varphi(\pi)$ belongs has nice proof-theoretic and model-theoretic properties. In particular, we can characterize those $\varphi(\pi)$ that are preserved under homomorphisms by modifying methods developed for the case of first-order logic by Craig and Lyndon. This leads at once to a resolution of the problem.

§1 Algorithmic formulas

By a relational structure $\underline{A}$ we understand a sequence $\underline{A} = \langle A; R_1, \ldots, R_m ; f_1, \ldots, f_n ; c_1, \ldots, c_k\rangle$ such that A is a non-empty set, R_i an m_i - ary relation on A, $i = 1, \ldots, m$, f_i an n_i-ary operation on A, $i = 1, \ldots, n$, and c_i an element of A, $i = 1, \ldots, k$. Two relational structures $\underline{A}$ and $\underline{B} = \langle B, S_1, \ldots, S_{m'} ; g_1, \ldots, g_{n'} ; d_1, \ldots, d_{k'}\rangle$ are similar if $m = m'$, $n = n'$, $k = k'$ and the ranks of the relations R_i and S_i and of the operations f_j and g_j are the same, $i = 1, \ldots, m$, $j = 1, \ldots, n$. The structure $\underline{B}$ is a homomorphic image of $\underline{A}$ if there is a map h of A onto B such that

$$\begin{cases} h(f_i(a_1, \ldots, a_{n_i})) = g_i(h(a_1), \ldots, h(a_{n_i})), & i = 1, \ldots, n, \\ R_i(a_1, \ldots, a_{m_i}) \text{ implies } S_i(h(a_1), \ldots, h(a_{m_i})), & i = 1, \ldots, m. \end{cases}$$

To each class of similar relational structures we associate a family of programs for constructions within the structures. These programs are interpreted under the assumption that decisions regarding equality and the relations can effectively be performed as atomic actions and that the basic operations can be executed. More precisely, distinguished elements:

Operational instructions

i : do $x_r := x_s$ then go to p;

i : do $x_r := f_j(x_{s_1},\ldots,x_{s_{n_j}})$ then go to p;

i : do $x_r := c_j$ then go to p

i : go to p

Conditional instructions

i : if $R_j(x_{s_1},\ldots,x_{s_{m_j}})$ then go to p else go to q .

The labels i, p, q are numerals.

Programs are finite sets of instructions in which no two different instructions have the same label and in which one instruction is singled out as the first instruction to be executed. An input to a program is an assignment of elements of a relational structure to the variables. For every input the program, together with the relational structure, determines a unique sequence of such assignments, called a computation, in the obvious manner. A program π terminates on an input if in the course of the execution a jump go to p has to be performed for a numeral p which is not the label of an instruction in π. Such a numeral will be called an exit of π.

Given a structure $\underline{A}$ and a program π we obtain, by the method developed in [2], a formula $\varphi(\pi)(x_0,x_1,\ldots)$ such that $\varphi(\pi)[a_0,a_1,\ldots]$ holds in $\underline{A}$ iff π terminates in $\underline{A}$ for the input $a_0, a_1,\ldots$. We don't need the details of this passage from programs to formulas with exception of the following facts which we present in tabular form:

Every program π is equivalent to a program in one of the following forms:	The corresponding formula is:
1 : <u>go to</u> 1 (denoted by "ν")	F (for "false")
1 : <u>if</u> $R_j(x_{s_1},\ldots,x_{s_{m_j}})$ <u>then go to</u> 2 <u>else go to</u> 1 (denoted by $\to\boxed{R_j}\to$) with $\downarrow\nu$	$R_j(x_{s_1},\ldots,x_{s_{m_j}})$
1 : if $R_j(x_{s_1},\ldots,x_{s_{m_j}})$ <u>then go to</u> 1 <u>else go to</u> 2 (denoted by $\to\boxed{R_j}\to\nu$) with $\downarrow$	$\neg R_j(x_{s_1},\ldots,x_{s_{m_j}})$
$\to\boxed{R_j}\to\pi_1$ with $\downarrow\nu$	$R_j(\ldots)\wedge\varphi(\pi_1)$
$\to\boxed{R_j}\to\nu$ with $\downarrow\pi_1$	$\neg R_j(\ldots)\wedge\varphi(\pi_1)$
$\to\boxed{x_r := f_j(x_{s_1},\ldots)}\to\pi_1$	$\mathrm{Sub}_{x_r}^{f_j(x_{s_1},\ldots)}(\varphi(\pi_1))$
For each input the computation according to π is the same as that according to π_1 or the same as that according to π_2	$\varphi(\pi_1)\vee\varphi(\pi_2)$

$\pi_0 \to \underset{\circlearrowleft}{\pi_1} \to \pi_2$ (π_0, π_2 may be missing)	$\bigvee_{k=0}^{\infty} \varphi^{(k)}$, where $\varphi^{(k)} = \varphi\left(\begin{array}{l} \pi_0 \to \pi_1 \to \pi_2 \\ \quad\;\; \downarrow \\ \left.\begin{array}{l} \pi_1 \to \pi_2 \\ \downarrow \\ \vdots \\ \downarrow \\ \pi_1 \to \pi_2 \end{array}\right\} k \text{ times} \\ \quad\;\; \downarrow \\ \quad\;\; \nu \end{array}\right)$

{ The last case is treated differently in [2], [3]. For the present applications it is important that subformulas of formulas $\varphi(\pi)$ are themselves of the form $\varphi(\pi')$; while for earlier applications other requirements where guiding the formulation.}

Formulas of the form $\varphi(\pi)$ are called algorithmic formulas; they are built up from atomic formulas by means of the syntactical operations described in the table above. Note that the syntactical operations $\neg, \vee$ and $\bigvee(\cdot)^{(k)}$ are only performable on certain types of algorithmic formulas.

Algorithmic formulas can be used to formulate algorithmic properties of structures and develop their theories, [1]). A different use is in connection with approximation theory, [2]). For the application that we have in mind in the present note we need to develop a free-variable deductive calculus of algorithmic formulas. We cast it in the Gentzen style using sequents.

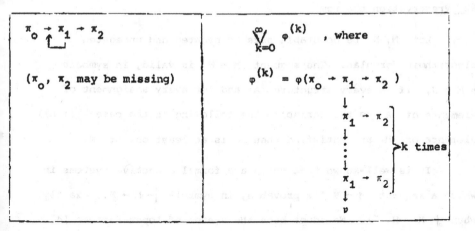

§2 Preservation theorem

Let M, N be countable sets of negated and unnegated algorithmic formulas. The sequent $M \to N$ is valid, in symbols $\models M \to N$, if in every structure $\underline{A}$ and for every assignment of elements of $\underline{A}$ to the variables the following is the case: If all elements of M are satisfied then so is at least one of N.

It is well-known that there are formal deductive systems in which a sequent $M \to N$ is provable, in symbols $\vdash M \to N$, exactly when $\models M \to N$, [3]). We adopt here the system of Lopez-Escobar [4] with some obvious modifications, namely each rule is formulated so that the formulas occuring in hypotheses and conclusion are all algorithmic (or negated algorithmic) formulas. For example, instead of a general rule

$$\frac{M \cup \{\varphi,\psi\} \to N}{M \cup \{\varphi\wedge\psi\} \to N}$$

we take the two rules

$$\frac{M \cup \{\alpha,\varphi\} \to N}{M \cup \{\alpha\wedge\varphi\} \to N}\ , \qquad \frac{M \to N \cup \{\neg\alpha,\neg\varphi\}}{M \to N \cup \{\neg(\alpha\wedge\varphi)\}}$$

where α is a negated or unnegated atomic formula. Similarly,

$$\frac{M \cup \{\varphi_k\} \to N \quad \text{for all} \quad k \in I,\ I \text{ countable}}{M \cup \{\bigvee_{k\in I} \varphi_k\} \to N}\ .$$

is modified to

$$\frac{M \cup \{\varphi^{(k)}\} \to N \text{ for } k = 0,1,\dots}{M \cup \{\bigvee_{k=0}^{\infty}\varphi^{(k)}\} \to N}\ ,\quad \frac{M \to N \cup \{\neg\varphi^{(k)}\} \text{ for } k = 0,1,\dots}{M \to N \cup \{\neg\bigvee_{k=0}^{\infty}\varphi^{(k)}\}}\ .$$

The rules obtained in this fashion obviously form a complete deductive system since the rules of [4] are complete.

If φ is any formula we let $Pos(\varphi)$ be the set of all predicate symbols occuring positively in φ and $Neg(\varphi)$ the set of all predicate symbols occuring negatively in φ : R occurs positively in $R(x_{s_1},\dots)$, negatively in $\neg R(x_{s_1},\dots)$;
$Pos(\varphi\vee\psi) = Pos(\varphi\wedge\psi) = Pos(\varphi) \cup Pos(\psi)$, $Pos(\bigvee_{k=0}^{\infty}\varphi^{(k)}) = \bigcup_{k=0}^{\infty} Pos(\varphi^{(k)})$,
$Neg(\varphi\vee\psi) = Neg(\varphi\wedge\psi) = Neg(\varphi) \cup Neg(\psi)$, etc.. Let $Var(\varphi)$ be the set of variables occuring in φ. A formula φ is called positive if $Neg(\varphi) = \emptyset$, the empty set, and negative if $Pos(\varphi) = \emptyset$. By abuse of notation we write $Pos(M)$ for $\bigcup_{\varphi\in M} Pos(\varphi)$, $Neg(M) = \bigcup_{\varphi\in M} Neg(\varphi)$.

<u>Interpolation theorem for algorithmic formulas</u>. Let M, N be countable sets of negated and unnegated algorithmic formulas. If $\vdash M \to N$ then there exists Q such that $\models M \to \{Q\}$ and $\models \{Q\} \to N$.
Moreover

(a) Q is a finite conjunction of negated and unnegated algorithmic formulas

(b) $Pos(Q) \subseteq Pos(M) \cap Pos(N)$, $Neg(Q) \subseteq Neg(M) \cap Neg(N)$, $Var(Q) \subseteq Var(M) \cap Var(N)$.

Our proof, similar to that of Craig [1] and Lopez-Escobar [4], is by induction on the depth of the proof of $M \to N$, 4). The interpolating formula of a conclusion of a rule is built up from the interpolating formulas of the hypotheses. This is considerably less straightforward than in [4] where arbitrary countable conjunctions and disjunctions are permissible, while we need to stay with algorithmic formulas. Thus, in particular, we need to carry along in the induction a proviso about the forms of interpolating formulas for sequents forming a family such as $M \cup \{\varphi^{(k)}\} \rightarrow N,\ k = 0,1,\ldots$.

Having obtained the interpolation theorem in this form, we can proceed to apply it for a proof of

Lyndon's theorem for algorithmic formulas: An algorithmic formula φ is preserved under homomorphisms iff φ is equivalent to a positive formula of the form $\varphi_1 \wedge \ldots \wedge \varphi_r \wedge \neg\psi_1 \wedge \ldots \wedge \neg\psi_s$ where φ_i, ψ_j, $i=1,\ldots,r$, $j = 1,\ldots,s$, are algorithmic formulas.

Our proof follows Lyndon's proof, [5], of the corresponding theorem for first-order logic with some modifications. Most of these are obvious. One essential application of the compactness theorem in Lyndon's proof (for corollaries 5.1, 5.2) can be replaced by another application of the interpolation theorem.

The above theorem solves the question posed in the introduction.

To treat the problem of preservation of programs in passage to actual computers there is still some refinement necessary, since the assumption of a homomorphism is only partially true, (only with respect to $<$, $\cdot$, $^{-1}$ but not in general with respect to $+$). A first tentative step in the direction of describing the relationship between ideal and actual computers by weakening the requirement of homomorphism has recently been taken by Rutishauser, (unpublished). It remains to find Lyndon's theorem for the kind of weak homomorphisms envisioned in that, or similar, proposals.

References

[1] W. Craig, Linear reasoning. A new form of the Herbrand-Geutzen theorem. Journal of Symbolic Logic 22 (1957), pp. 250 - 268.

[2] E. Engeler, Algorithmic properties of structures. Mathematical Systems Theory 1 (1967), pp. 183 - 195.

[3] E. Engeler, Formal Languages: Automata and Structures, Markham Publ. Co., Chicago, 1968.

[4] E. G. K. Lopez-Escobar, An interpolation theorem for denumerably long formulas. Fundamenta Mathematicae 52 (1965), pp. 253 - 272.

[5] R. C. Lyndon, Properties preserved under homomorphisms. Pacific Journal of Mathematics 9 (1959), pp. 143 - 154.

Footnotes

*) The present note reports on results obtained with partial support under NSF Grant GP 5434; a more detailed version with proofs forms part of a forthcoming larger project.

1) In particular the algorithmic theories of natural numbers, integers and rational numbers, [3], and the algorithmic theories of real numbers and geometry, in [Remarks on the theory of geometrical constructions. The Syntax and Semantics of infinitary Languages (J. Barwise, ed.) Lecture Notes in Mathematics (Springer), 72 (1968), pp. 64 - 76.]

2) Approximation by Machines, Part I, 26 pp., mimographed Forschungsinstitut f. Mathematik, ETH Zürich, 1968.

3) For example by Carol Karp, by Lopez-Escobar [4], and in the following two papers by the author: Zur Beweistheorie von Sprachen mit unendlich langen Formeln. Zeitschrift f. math. Logik, Grundl. d. Math. 7 (1961). pp. 213 - 218; A reduction principle for infinite formulas, Mathematische Annalen 151 (1963), pp. 296 - 303. The earlier system could in fact be used to prove the interpolation theorem, while the later apparently does not lend itself to it, as remarked in [4].

4) There are also model-theoretic proofs for the first-order case (by Lyndon, Henkin and Keisler.) Kreisel, Barwise and others have far-reaching results for certain other fragments of $L_{\omega_1\omega}$; we have not yet established the connection to that work.

ASPECTS DU THEOREME DE COMPLETUDE SELON HERBRAND,

Roland FRAÏSSE

L'étude des déployées herbrandiennes est l'une des approches les plus utilisables en démonstration automatique.

Un problème préalable est la construction d'une formule qui soit une thèse si et seulement si la question mathématique posée se résoud affirmativement. Prenons l'exemple classique de l'énoncé de Fernat, dans le cadre de l'axiomatique de l'arithmétique selon Peano, munie des axiomes usuels de l'addition et de la multiplication, et complétée par les axiomes récurrents de l'exponentielle, dont on sait qu'ils sont non créatifs. On pourra rendre finie cette axiomatique, donc la réduire à un seul axiome, par la méthode des prédicats additionnels (Kleene, 1952, Craig et Vaught, 1958).

Ce préalable étant assuré, on peut présenter comme suit les problèmes théoriques en rapport avec la programmation d'une machine (vers une solution positive de l'énoncé de Fermat, par exemple, la programmation usuelle des machines à calculer n'étant évidemment adaptée qu'à une solution négative).

1 - *Thèse libre.*

Rappelons *l'algorithme classique décidant si une formule libre* (sans $\forall$ ni $\exists$) *est une thèse.* Numérotant les individus x_i $(i = 1,2,\dots)$ on remplacera de toutes les façons chaque x_i par un entier $\leq i$. Chaque identité ainsi obtenue entre deux entiers sera remplacée par sa valeur de vérité, et chaque formule atomique formée d'un prédicat suivi d'entiers sera remplacée par un atome du calcul des connections. Exemple:

$\rho\, x_1\, x_1 \vee \neg\, \rho\, x_1\, x_2 \vee (\rho\, x_1\, x_2 \wedge x_1 \neq x_2)$: remplacer x_1 par 1 et x_2 par 2 , et $1 \neq 2$ par la valeur "vrai"; puis remplacer x_1 et x_2 par 1 et $1 \neq 1$ par "faux".

2 - *Thèse explicite.*

Etant donnée une formule logique (avec ses quanteurs $\forall$ et $\exists$) la qu stion se pose de l'écrire ou non sous forme prénexe; des arguments divergents ont été présentés sur ce choix. Adoptons la forme prénexe pour commencer; nous étudierons plus loin (paragraphe 9) le cas non prénexe.

Appelons ***thèse explicite*** une formule prénexe dans laquelle on peut remplacer chaque $\exists$-individu par un individu qui le précède dans le préfixe, ou par un individu libre, de façon à obtenir une thèse libre. Le nombre d'essais étant fini, la notion de thèse explicite est décidable. Exemple : $\forall_2\ \exists_1 (\neg\, \rho\, x_1 \vee \rho\, x_2)$ est une thèse explicite (remplacer 1 par 2) alors qu'en écrivant le préfixe $\exists_1\ \forall_2$, la thèse obtenue n'est plus explicite. Dans le cas général d'une formule en $\neg$, $\wedge$, $\vee$, il n'est pas possible d'obtenir une thèse explicite à partir d'une thèse donnée, par simple déplacement des quanteurs. Par exemple la thèse $(\forall_1\ \rho\, x_1 \wedge \forall_2\ \rho\, x_2) \vee \exists_3 \neg\, \rho\, x_3$ ne devient jamais explicite, que l'on sorte les quanteurs dans l'ordre $\forall_1\ \forall_2\ \exists_3$ ou dans tout autre ordre.

3 - *Déployée dure.*

Nous allons définir les ***déployées*** d'une formule donnée P . Ce seront des formules *équidéduites* de P , c'est à dire prenant la même valeur pour chaque suite de relations remplaçant les prédicats, et d'éléments de la base remplaçant les individus libres. Le principal intérêt de la déployée résulte du *théorème de complétude* selon (Herbrand 1930, 1931), voir aussi (Gentzen, 1934).

Nous le présenterons ainsi: une *formule prénexe est une thèse si et seulement s'il existe une thèse explicite parmi ses déployées.*

Deux notions au moins sont possibles: la déployée *dure* ou *souple*. Notons x_1 à x_n les individus libres, x_{n+1} à x_{n+p} les individus liés, dans l'ordre du préfixe. Donnons-nous une suite finie de fonctions f_i , définies sur les entiers 1 à $n+p$, avec $f_i(x) = x$ pour $x = 1,\ldots,n$ et chaque f_i croissante. Pour la première $f_1(x) = x$ pour $x = 1,\ldots,n+p$. A chaque $f_i\ (i \geq 2)$ associons un *numéro de renvoi* $u(i) < i$ et un *rang de reprise* $r(i)$, avec $n+1 \leq r(i) \leq n+p$; il sera commode de poser $r(1) = n+1$.

Posons $f_i(x) = f_{u(i)}(x)$ pour $x = 1,\ldots,n,\ldots,r(i) - 1$; les valeurs pour $x = r(i),\ldots,n+p$ étant, dans l'ordre croissant, les premiers entiers non encore utilisés comme valeurs de $f_1,\ldots,f_{i-1}$. Notons U le préfixe, A la partie libre de la formule donnée; *l'assignée* A_i sera la transformée de A par f_i (notons que $A_1 = A$); le *préfixe repris* U_i sera le transformé par f_i de l'intervalle final de U, allant du rang de reprise $r(i)$ au rang final $n+p$ (notons que $U_1 = U$). Alors la déployée dure sera définie comme $U_1\, U_2 \ldots U_h\, (A_1 \vee A_2 \vee \ldots \vee A_h)$, où h est le nombre des f_i, ou des assignées. Une telle déployée sera dite *h ème*, ou *d'ordre* h.

Exemple. La formule suivante :

$\exists_1\, \forall_2\, \exists_3 \neg \rho\, x_1\, x_2 \vee (\rho\, x_2\, x_3 \wedge x_2 \neq x_3) \vee \rho\, x_2\, x_2$ est une thèse; on obtient comme thèse explicite la déployée troisième dont les préfixes sont

$\exists_1\, \forall_2\, \exists_3$; $\exists_4\, \forall_5\, \exists_6$ (renvoi 1, reprise 1); $\exists_7$ (renvoi 1, reprise 3)

donc les deuxième et troisième assignées s'écrivent :

$$\exists_4\, \forall_5\, \exists_6 \neg \rho\, x_4\, x_5 \vee (\rho\, x_5\, x_6 \wedge x_5 \neq x_6) \vee \rho\, x_5\, x_5$$

$$\exists_7 \neg \rho\, x_1\, x_2 \vee (\rho\, x_2\, x_7 \wedge x_2 \neq x_7) \vee \rho\, x_2\, x_2$$

Remplaçant 4 par 2 et 7 par 5, on obtient la thèse libre...

$\vee\, \rho\, x_2\, x_2 \vee \neg \rho\, x_2\, x_5 \vee \ldots \vee (\rho\, x_2\, x_5 \wedge x_2 \neq x_5) \vee \ldots$

4 - *Equidéduction avec la formule de départ.*

On peut toujours choisir le numéro de renvoi $u(i)$ le plus petit possible pour chaque i : alors le rang de reprise de la *i ème* assignée est soit égal à $n+1$, soit strictement supérieur au rang de reprise de la $u(i)$ *ème* assignée.

Soit alors une déployée *h ème*, et soit $k = u(h)$ le numéro de renvoi de h ; écrivons le préfixe U_k sous la forme $V\,W$, où V correspond aux rangs $r(k)$, $r(k)+1,\ldots, r(h)-1$ et W aux rangs $r(h),\ldots, r(n+p)$ qui sont ceux de U_h. Alors $W\,A_k$ est équidéduite de $U_h\,A_h$ et la déployée *h-ème* considérée prend la forme $U_1 \ldots U_{k-1} V\,W\,B$ avec B équidéduite de :

$$A_1 \vee \ldots \vee A_k \vee U_{k-1} \ldots U_{h-1}\, (A_{k+1} \vee \ldots \vee A_{h-1} \vee W\, A_k)$$

Les indices de $U_{k+1},\ldots, U_{h-1}$ ne figurant pas dans A_k, on peut encore remplacer B par :

$$A_1 \vee \ldots \vee A_k \vee W\, A_k \vee U_{k+1} \ldots U_{h-1}\ (A_{k+1} \vee \ldots \vee A_{h-1})$$

Utilisons alors les contractions classiques suivantes, où W désigne une suite de quanteurs et R, S deux relations : $W(R \vee W\,R)$ équidéduite de $W\,R$, puisque les indices de W sont inactifs dans $W\,R$; par suite $W(R\ V\ S \vee W\,R)$ équidéduite de $W(R \vee S)$. On obtient à la place de $W\,B$ la formule équidéduite contractée :

$W(A_1 \vee \ldots \vee A_k \vee U_{k+1} \ldots U_{h-1}\ (A_{k+1} \vee \ldots \vee A_{h-1})$ et par conséquent la déployée h-1-*ème* à la place de la déployée h-*ème*.

De ce qui précède, il résulte que l'existence d'une thèse explicite parmi les déployées de P entraine que P est une thèse. La réciproque est plus difficile à obtenir; voir (Herbrand 1930) ou encore (Henkin 1949), (Craig 1957), (Fraïssé 1967 Ch. 8).

5 - *Déployée souple.*

Elle s'obtient à partir de la précédente en changeant arbitrairement l'ordre des quanteurs dans le préfixe, pourvu que l'ordre soit conservé dans le préfixe restreint à chaque assignée. Avec l'exemple, que nous restreignons à la déployée seconde (indices 1 à 6), récrivons le préfixe dans l'ordre nouveau $\exists_1\ \forall_2\ \exists_4\ \forall_5\ \exists_3\ \exists_6$, avec encore les substitutions 1, 2, 3 et 4, 5, 6 ;

On obtient :

$$\begin{aligned}\exists_1\ \forall_2\ \exists_4\ \forall_5\ \exists_3\ \exists_6\ \neg\,\rho\, x_1\, x_2\ &\vee(\rho\, x_2\, x_3 \wedge x_2 \neq x_3) \vee \rho\, x_2\, x_2\\ \vee\ \neg\,\rho\, x_4\, x_5\ &\vee(\rho\, x_5\, x_6 \wedge x_5 \neq x_6) \vee \rho\, x_5\, x_5\end{aligned}$$

où il suffit de remplacer 4 par 2 et 3 par 5 pour avoir une thèse libre.

On peut, de façon équivalente, définir la déployée souple directement sans ordonner l'ensemble des assignées, en se donnant seulement les conditions suivantes. Chaque assignée A est encore la transformée de A par une fonction croissante f_i définie sur les rangs $1, 2, \ldots, n+p$ avec encore $f(1) = 1, \ldots\ f(1) = 1, \ldots, f(n) = n < f(n+1) < \ldots < f(n+p)$. La seule condition à satisfaire par les f est celle-ci :

Lorsque deux fonctions f prennent une même valeur, ce ne peut être qu'au même rang x, et alors elles prennent la même valeur à chaque rang $\leqslant x$. Pour chaque assignée, nous définirons son *préfixe associé*, comprenant les valeurs de f pour chacun des rangs $n+1$ à $n+p$; le préfixe général est astreint seulement à respecter l'ordre des quanteurs dans chaque préfixe associé. Dans l'exemple ci-dessus, les préfixes associés sont $\exists_1 \forall_2 \exists_3$ et $\exists_4 \forall_5 \exists_6$

Puisque la déployée dure peut être considérée comme une déployée souple particulière, le théorème de complétude subsiste avec les déployées souples. On peut même exiger, en déployée souple, que le numéro de renvoi soit toujours $u(i) = i-1$. Il suffit d'ordonner partiellement les valeurs prises par les fonctions f_i dans la déployée souple, en considérant $f_i(x)$ antérieure à $f_j(x')$ lorsque $x \leqslant x'$ et $f_j(x) = f_i(x)$. L'ordre partiel obtenu est un arbre fini; or dans un arbre fini on peut numéroter les restrictions totalement ordonnées maximales ou *chaînes maximales* pour l'inclusion, de sorte que chaque chaîne maximale reproduise la précédente jusqu'à un certain élément, puis soit formée d'éléments nouveaux, c'est à dire n'appartenant à aucune des chaînes déjà numérotées.

6 - *Rang de reprise et numéro de renvoi différents de* 1 .

Il est facile de donner un exemple où la résolution de la thèse exige un rang de reprise autre que 1 . (nous lions tous les individus, donc $n=0$ et les individus liés sont notés $x_1,\dots,x_p$) . Cela aussi bien en déployée souple qu'en déployée dure; prenons :

$$\forall_1 \forall_2 \exists_3 (\rho\ w_1\, x_1 \wedge \rho\, x_1\, x_2) \vee \neg \rho\, x_1\, x_3$$

Il faudra nécessairement faire quelque part la reprise au rang 2 ou 3, obtenant par exemple la thèse explicite :

$$\forall_1 \forall_2 \exists_3 \exists_4 (\rho\, x_1\, x_1 \wedge \rho\, x_1\, x_2) \vee \neg \rho\, x_1\, x_3 \vee \dots \vee \neg \rho\, x_1\, x_4 .$$

Voici un exemple, communiqué par J.P. Bénéjam, d'une thèse exigeant au moins un numéro de renvoi différent de 1 , en déployée dure :

$$\forall_1 \exists_2 \forall_3 \exists_4 (\rho\, x_1\, x_2 \wedge \rho\, x_3\, x_3 \wedge \rho\, x_2\, x_4) \vee \neg \rho\, x_4\, x_5$$

(on peut prendre comme thèse explicite la déployée dure troisième suivante, comportant le renvoi $u(2)=1$ (avec rang de reprise $r(2)=2$) et le renvoi

$u(3) = 2$ (rang de reprise $r(3) = 4$) :

$$\forall_1 \exists_2 \forall_3 \exists_4 (\rho x_1 x_2 \wedge \rho x_3 x_3 \wedge \rho x_2 x_4) \vee \neg \rho x_4 x_3$$

$$\exists_5 \forall_6 \exists_7 \vee(\rho x_1 x_5 \wedge \rho x_6 x_6 \wedge \rho x_5 x_7) \vee \neg \rho x_7 x_6$$

$$\exists_8 \vee(\rho x_1 x_5 \wedge \rho x_6 x_6 \wedge \rho x_5 x_8) \vee \neg \rho x_8 x_6$$

La résolution s'obtient en remplaçant 4 par 1, 5 et 8 par 3, 7 par 6 (parenthèse de la deuxième assignée compensée par les termes en $\neg \rho$).

Raisonnons dans le cas de la déployée dure. Si tous les renvois se font à la première assignée, compte-tenu de ce que les reprises ne sont utiles qu'aux deuxième et quatrième rangs, on obtient des parenthèses des formes suivantes :

$$\rho x_1 x'_2 \wedge \rho x'_3 x'_3 \wedge \rho x'_2 x'_4 \quad \text{ou} \quad \rho x_1 x_2 \wedge \rho x_3 x_3 \wedge \rho x_2 x'_4$$

Dans le deuxième cas, le terme $\rho x_1 x_2$ ne peut pas être compensé. Dans le premier cas, $\rho x'_3 x'_3$ est compensé par le terme $\neg \rho x'_4 x'_3$ de la même assignée : sinon ce serait par le terme $\neg \rho x''_4 x'_3$ d'une assignée ne renvoyant pas à la première. Alors le terme $\rho x'_2 x'_4$ prend la forme $\rho y x'_3$ où $y \neq x'_3$; il ne peut plus être compensé que par le terme $\neg \rho x''_4 x'_3$ d'une assignée ne renvoyant pas à la première.

Notons qu'en déployée souple on obtient la thèse explicite suivante, où les deux renvois se font à la première assignée :

$$\forall_1 \exists_2 \forall_3 \exists_4 (\rho x_1 x_2 \wedge \rho x_3 x_3 \wedge \rho x_2 x_4) \vee \neg \rho x_4 x_3$$

$$\exists_5 \forall_6 \exists_7 \vee(\rho x_1 x_5 \wedge \rho x_6 x_6 \wedge \rho x_5 x_7) \vee \neg \rho x_7 x_6$$

$$\exists_8 \vee(\rho x_1 x_2 \wedge \rho x_3 x_3 \wedge \rho x_2 x_8) \vee \neg \rho x_8 x_3$$

On peut prendre le préfixe $\forall_1 \exists_5 \forall_6 \exists_2 \forall_3 \exists_4 \exists_7 \exists_8$ et remplacer 2 et 4 par 6, 7 par 1, 8 par 3 (parenthèse de la 3ème assignée compensée par les $\neg \rho$) .

Par une légère complication de cet exemple, communiqué également par J.P. Bénéjam, on est obligé même en déployée souple, de se donner au moins un renvoi différent de 1 ; il s'agit de la thèse :

$$\forall_1 \exists_2 \forall_3 \exists_4 (\rho x_1 x_2 \wedge \rho x_3 x_3 \wedge \rho x_2 x_4 \wedge \sigma x_2 x_2 \wedge \sigma x_2 x_3) \wedge \neg \rho x_4 x_3 \wedge \neg \sigma x_3 x_4$$

On obtient une thèse explicite, avec une deuxième assignée (reprise 2) une troisième et une quatrième assignées (renvoi 1, reprise 4), enfin une

La première idée, qui vient pour une formule en $\neg$, $\wedge$, $\vee$, est d'accorder systématiquement la priorité au quanteur qui donnera $\forall$ sur celui qui donnera $\exists$, parmi les sortants immédiats. Par exemple, à partir de $\forall_1\, \rho\, x_1 \vee \neg \forall_2\, \rho\, x_2$, on fait sortir d'abord $\forall_1$ qui reste $\forall$, puis $\forall_2$ qui donne $\exists$. Dans le cas où tous les sortants immédiats sont tous $\forall$ ou tous $\exists$, on pourrait faire sortir d'abord le quanteur dont la sous-formule correspondante est la plus courte, avec une règle précisant le cas de deux formules également courtes. Or cela parait une mauvaise stratégie. Par exemple partons de :

$$\exists_1\, \exists_2\, \forall_3 (x_1 = x_3 \vee x_2 = x_3 \vee \rho\, x_3)$$

$$\exists_4 (\neg \rho\, x_4 \wedge \forall_5\, x_4 = x_5) \vee \exists_6\, \exists_7\, x_6 \neq x_7 .$$

La stratégie proposée ferait sortir les quanteurs dans l'ordre 6, 7, 4, 5, 1, 2, 3 ce qui ne conduit pas à une thèse explicite ; alors que l'ordre 1, 2, 3, 4, 5, 6, 7 y conduit.

9 - *Tentatives pour le cas non prénexe.*

Revenons sur le choix fait au paragraphe 2, pour adopter ici la forme générale, non forcément prénexe, de la formule que l'on soupçonne être une thèse.

Si l'on ne peut pas traiter les formules générales comme des formules prénexes déguisées, la première précaution à prendre est de traiter directement, entre autres, les *tautologies*. Il s'agit des formules formées à l'origine des connections $\neg$, $\wedge$, $\vee$ et d'atomes, et qui prennent la valeur vrai quelles que soient les valeurs, vrai ou faux, attribuées aux atomes; ces derniers étant remplacés chacun par une formule logique. Par exemple il nous faut reconnaitre d'emblée que $\forall_x\, \rho\, x \vee \neg \forall_x\, \rho\, x$ est une thèse, sans avoir à changer un x en y, ni à remplacer $\neg \forall_y$ par $\exists_y \neg$, ni remplacer enfin y par x. Il faut donc au départ étendre la notion de *thèse libre* de façon à y englober à la fois les thèses libres et les tautologies. Partant d'une thèse libre, proposons-nous de la conserver à l'état de thèse en lui ajoutant des quanteurs convenablement répartis. Par exemple autorisons le passage de la thèse $\rho\, x \vee \neg \rho\, x$ à la thèse $\forall_x\, \rho\, x \vee \neg \forall_x\, \rho\, x$, mais interdisons-nous le passage à la non-thèse cinquième assignée (renvoi 2, reprise 4) ; ce qui donne :

$$\exists_5 \forall_6 \exists_7 (\rho\, x_1 x_5 \wedge \rho\, x_6 x_6 \wedge \rho\, x_5 x_7 \wedge \sigma\, x_5 x_5 \wedge \sigma\, x_5\, x_6) \vee \neg\rho\, x_7\, x_6 \vee \neg\sigma\, x_6\, x_7$$

$\exists_8$ renvoi 1ère assignée, donc $\ldots \vee \neg\rho\, x_8\, x_3 \vee \neg\sigma x_3\, x_8$

$\exists_9$ renvoi 1ère assignée, donc $\ldots \vee \neg\rho\, x_9\, x_3 \vee \neg\sigma x_3\, x_9$

$\exists_{10}$ renvoi 2ème assignée, donc $\ldots \vee \neg\rho\, x_{10}\, x_6 \vee \neg\sigma\, x_6\, x_{10}$

Remplacer 4 par 1; 5, 8, 10 par 3; 7, 9 par 6 (parenthèse 2ème assignée compensée par les $\neg\rho$ et $\neg\sigma$).

7 - *Sur la réduite herbrandienne utilisant les fonctions skolemiennes.*

La déployée, dure ou souple, peut être considérée comme un allègement de la *réduite* classique, utilisant les fonctions de (Skolen,1920). Par exemple, partant de la thèse des paragraphes 3 et 5, cette réduite s'obtient en remplaçant d'abord chaque $\forall$-individu par une fonction des $\exists$-individus qui le précèdent dans le préfixe. Donc on remplace 2 par une fonction $s(1)$ et 5 par une fonction $t(1,4)$, obtenant la formule :

$$\neg\rho\, x_1\, x_{s1} \vee (\rho\, x_{s1}\, x_3 \wedge x_{s1} \neq x_3) \vee \rho\, x_{s1}\, x_{s1}$$

$$\neg\rho\, x_4\, x_{t1,4} \vee (\rho\, x_{t1,4}\, x_6 \wedge x_{t1,4} \neq x_6) \vee \rho\, x_{t1,4}\, x_{t1,4}$$

où il est entendu que 1, 3, 4, 6 sont tous sous quanteur $\exists$. Il est évidemment nécessaire, comme auparavant, de substituer par exemple 4 par $2 = s(1)$ et 3 par $5 = t(1,s(1))$, ce qui donne enfin une thèse libre : En somme, les fonctions skolémiennes, qui ont par ailleurs un grand intérêt théorique, ne servent ici qu'à surchager le travail de résolution des thèses.

8 - *Sur la stratégie du déplacement des quanteurs.*

Partons d'une thèse logique P non prénexe, et demandons-nous quelle est la meilleure stratégie pour transformer P en une thèse prénexe par déplacement des quanteurs. Le meilleur résultat serait une thèse explicite lorsque c'est possible; sinon une thèse prénexe admettant une déployée du plus petit ordre possible, qui soit thèse explicite.

$\forall_x \; \rho \, x \vee \forall_x \neg \quad x$. La première idée qui vient est d'exiger que les sous-formules crées par l'apparition du nouveau quanteur et débutant par chacune de ses occurrences, soient identiques. C'est bien l'identité qu'il nous faut. On ne saurait se contenter de formules transformées les unes des autres par permutation des individus libres; ce serait autoriser le passage de la thèse $x \neq y \vee x \neq z \vee y = z$ à la non-thèse $(\forall_x \, x \neq y) \vee (\forall_x \, x \neq z) \vee y = z$. De plus nous exigerons que le nouveau quanteur ne soit pas dominé, pour éviter par exemple le passage de la thèse $\exists_y \; x = y$ à la non-thèse $\exists_y \forall_x \, x = y$.

Appelons donc *thèse pseudo-libre* toute formule obtenue d'une thèse libre en ajoutant un nombre fini de fois, un quanteur non dominé, toutes les sous-formules débutant par ses diverses occurrences étant identiques, et un même individu n'ayant pas à la fois des occurrences libres et liées.

Cas particuliers. Ces thèses libres, éventuellement précédées d'un préfixe arbitraire; les tautologies : on passera de $\rho \, x \, y \vee \neg \rho \, x \, y$ à $\exists_y \rho \, x \, y \vee \neg \exists_y \rho x \, y$ puis à $\forall_x \exists_y \rho \, x \, y \vee \neg \forall_x \exists_y \rho \, x \, y$; enfin les formules qui n'appartiennent pas aux deux sortes précédentes, comme la formule:

$$\forall_x \; \rho \, x \vee (\neg \forall_x \; \rho \, x \wedge y = z) \vee \; y \neq t \vee z \neq t \; .$$

Pour voir que toute thèse pseudo-libre est bien une thèse, raisonnons dans le cas simple d'une formule logique $A \, x \, y \, z$ ayant pour individus libres x, y, z et se présentant sous deux occurrences. Alors la formule à examiner prend par exemple la forme $\Phi_{y\,z}(A x \, y z) \vee \Psi_{yz}(A x \, y z)$, où $\Phi_{yz}(a)$ est une formule connective dont un atome a est encore libre, les autres ayant été remplacés par des formules logiques des individus libres y, z .

Supposons que la formule donnée est une thèse, et montrons qu'il en est de même de sa transformée par addition de deux occurrences du quanteur $\exists_x$, par exemple :

$\Phi_{yz}(\exists \;\; A x \, y z) \vee \Psi_{yz}(\exists \;\; A x \, y z)$. Si cela est faux, il existe des relations pour les prédicats, un élément pour y et un pour z rendant faux $\Phi_{yz}(\exists \;\; A x \, y z)$ et $\Psi_{yz}(\exists \;\; A x \, y z)$. Ou bien $A x \, y z$ prend la même valeur de vérité pour tout x , le y et le z précédents. Alors cet y, ce z et tout x rendent faux $\Phi_{yz}(A x \, y z)$ et $\Psi_{yz}(A x \, y z)$ donc $(\Phi \vee \Psi)(A)$ n'est pas une thèse. Ou bien cet y , ce z et un x au moins donnent à $A x \, y z$ la valeur vrai, un autre x au moins donnant la valeur faux. Alors

$\exists_x A\,x\,y\,z$ prend la valeur vrai, Φ_{yz} (vrai) et Ψ_{yz} (vrai) prennent la valeur faux. Finalement cet y , ce z et un x au moins donnent à $A\,x\,y\,z$ la valeur vrai donc la valeur faux à $\Phi_{yz}\,(A\,x\,y\,z)$ et à $\Psi_{yz}\,(A\,x\,y\,z)$ donc à $(\Phi \vee \Psi)\,(A)$ qui n'est pas une thèse.

L'algorithme de décision pour les thèses pseudo-libres est immédiat; on supprime tous les quanteurs, on vérifie qu'il reste une thèse libre, puis on cherche à rétablir les quanteurs un à un en respectant la définition.

Proposons-nous maintenant d'étendre la notion de *thèse explicite*; nous appellerons *thèse pseudo explicite* toute formule qui donne une thèse pseudo-libre par un nombre fini d'opérations suivantes. Diversification des quanteurs d'une même sous-formule, rassemblement en une suite, dite préfixe local, en tête de cette sous-formule. Les connections étant $\neg$, $\wedge$, $\vee$, ce préfixe local se trouve en position positive ou négative selon qu'il est dominé par un nombre pair ou impair de négations. Dans le premier cas, substitution de certains $\exists$- individus par d'autres qui les précèdent dans le préfixe, ou par des individus libres de la sous-formule. Dans le second cas, substitution analogue de certains $\forall$- individus. Redéplacement des quanteurs et réunification d'occurrences lorsque cela respecte la condition que les sous-formules débutant par elles soient identiques. Ces opérations ne changent pas la superposition des connections, identités et prédicats, sont en nombre fini; on a donc un algorithme de décision pour les thèses pseudo-explicites. D'autre part elles englobent évidemment les thèses explicites.

Pour que le théorème de complétude s'étende, il ne reste plus qu'à préciser la notion de *pseudo-déployée*. Il suffit, semble-t-il, de reprendre l'opération déjà considérée de constitution d'un préfixe local U , mais au lieu de substituer des individus, on démultiplie U et la formule qui le suit comme il est fait, par exemple, à l'aide des fonctions f du paragraphe 3, et on remplace $U\,A$ indifféremment par $U_1 \ldots U_h\,(A \quad \vee \ldots \vee A_h)$ ou par $U_1 \ldots U_h\,(A_1 \wedge \ldots \wedge A_h)$ le $\vee$ se révélant commode dans le cas d'une occurrence positive, et le $\wedge$ dans le cas d'une occurrence négative. Après quoi on s'autorise à remplacer les quanteurs.

Tout ce qui est dit au paragraphe 4 subsiste, même si les A sont des formules logiques avec quanteurs : on a donc l'équidéduction voulue. Enfin ces manipulations comprenant en particulier celles du cas prénexe, toute thèse admet au moins une pseudo-déployée, à savoir une déployée prénexe, qui est une thèse explicite, donc une thèse pseudo-explicite. Le théorème de complétude subsiste donc. D'un côté on a une plus grande souplesse, ne serait-ce que par la possibilité de traiter immédiatement les tautologies. Mais dans un traitement mécanique, celane sera-t-il pas compensé par la variété plus grande des déployées.

10 - *Considérations historiques.*

Ce n'est pas méconnaître l'importance de Herbrand (ni se diminuer soi-même) que de reconnaître dans sa rédaction de nombreuses obscurités ou d'y découvrir des lemmes faux. (Aanderaa, Andrews et Dreben, 1963). Citons à ce sujet la préface de (Herbrand, 1968) rédigée par J. Van Heijenoort, p. 11 et 12. "De nombreux passages dans ces écrits portent les marques d'une rédaction hâtive. Les textes contiennent, outre les erreurs de raisonnement que nous avons discutées plus haut, des difficultés de tous genres coquilles (...) renvois (à des pages, des paragraphes) inexacts, phrases grammaticalement mal construites et par conséquent ambiguës, ponctuation capricieuse, mots qui manquent, enfin passages obscurs qui demanderaient bien des éclaircissements (...) Il existe du chapitre 5 de la thèse de Herbrand, qui est peut-être le texte présentant le plus de difficultés, une traduction anglaise par Burton Dreben et moi-même, dans Van Heijenoort 1967, p. 525-581. Les traducteurs se sont efforcés de rendre en anglais de nombreux passages moins obscurs qu'ils ne sont dans l'original, et le lecteur, même de langue française, trouvera peut-être quelque profit à consulter cette traduction". (fin de citation).

On ne s'étonnera pas s'il nous parait difficile de savoir avec précision lesquelles, parmi les notions précédemment considérées, figurent déjà chez Herbrand. La thèse explicite figure sous le nom d'identité normale (1930, Ch. 5, n° 2-1) ; la déployée souple parait être introduite directement sous le nom de proposition dérivée (Ibidem, n° 2-2). Nous avons renoncé à trouver la trace de la déployée dure. Il est en tous cas plus facile de la réinventer que de déchiffrer les paragraphes qui pourraient éventuellement la concerner. Enfin la déployée non prénexe existe chez Herbrand mais nous n'avons pu discerner comment elle se résoud, notamment si on conserve des occurrences distinctes pour le même quanteur, et selon quelles règles. Or nous avons vu au paragraphe 9 quelles difficultés soulèvent de telles règles, et d'autre part si l'on rend distincts les quanteurs des occurrences distinctes, alors le travail de résolution se calque sur celui du cas prénexe, et tout bénéfice, par rapport à ce cas, disparait.

Nous espérons que le présent article, qui reprend avec plus de détails celui de Bénéjam et Fraïssé, 1964, aura permis au lecteur pressé de se faire une idée du théorème de complétude, et donnera au lecteur patient l'envie de s'attaquer à la bibliographie, Herbrand inclus.

BIBLIOGRAPHIE

S. AANDERAA, P. ANDREWS et B. DREBEN, 1963 - False lemmas in Herbrand - Bulletin of the Amer. Math. Soc. t.69, n°5, p.699-706.

J.P. BÉNÉJAM et R. FRAÏSSÉ, 1964 - Sur le théorème de Herbrand-Gentzen, Gauthier-Villars.

W. CRAIG, 1957 - Linear reasoning - A new form of the Herbrand-Gentzen theorem Journal Symbolic logic, t.22, p. 250-268.

W. CRAIG et VAUGHT, 1968 - Finite axiomatizability using additional predicates - Journal of symbolic logic, t.23, p. 289-308.

R. FRAÏSSÉ, 1967 - Cours de logique mathématique, t.1, p. 186, Paris Gauthier-Villars.

G. GENTZEN, 1934 - Untersuchungen über das logische schliessen. Math. Zeitschrift t.39, p. 176-210 et 405-431. Traduction française : Recherches sur la déduction logique par R. Freys et J. Ladrière, 1955, Paris. (Presses Universitaires de France, 170 p.).

J. VAN HEIJENOORT, 1967 - From Frege to Gödel, Harvard University Press.

L. HENKIN, 1949 - The completeness of the first order functional calculus - Journal of symbolic logic, t.14, p. 159-166.

J. HERBRAND, 1930 - Recherches sur la théorie de la démonstration (thèse) Soc. sciences et lettres, Varsovie, Cl.3 (math. phys.) n°33.

1931 - Sur le problème fondamental de la logique mathématique - Comptes rendus Soc. des sciences et lettres, Varsovie, Classe 3, n°24.

1968 - Ecrits logiques (J. Van Heijenoort, éditeur) Presses Universitaires de France, Paris.

S.C. KLEENE, 1952 - Finite axiomatizability of theories in the predicate calculus using additional predicate symbols - Memoirs of the Amer. Math. soc. n° 10, p. 27-68.

J. LOS, A. MOSTOWSKI et H. RASIOWA, 1956 - A proof of Herbrand theorem - Journal de math. pures et appl. t. 35, p. 19-24.

R.C. LYNDON, 1959 - An interpolation theorem in the predicate calculus - Pacific Journal of math. t. 9, p. 129-142.

W.V. QUINE, 1955 - A proof procedure for quantification theory - Journal of symbolic logic, t. 20, p. 141-149.

T. SKOLEM, 1920 - Logisch-kombinatorische untersuchungen über die Erfullbarkeit oder Beweisbarkeit mathematischer sätzenebst einem theoreme über dichte Mengen - Skrift Videnskap. Kristiania, I (nat. - natur) n° 4, 36 pages.

H. WANG, 1962 - A survey of mathematical logic, Pékin.

DECISION PROCEDURE FOR THEORIES CATEGORICAL IN ALEF$_0$

Andrzej Grzegorczyk

The decision procedure for categorical theories suggested by well known test of R.L.Vaught [5] is not the simplest one. In many cases a simpler procedure may be obtained using a kind of elimination of quantifiers which is universal for theories categorical in alef$_0$. We start by study of models for theories categorical in alef$_0$. Then we pass to the decision procedure. At the end we make some suggestions concerning the interesting conjecture of decidability of every categorical theory having finite number of primitive notions.

1. Models categorical in alef$_0$

Categorical theories are given rather by looking at their models like dense ordering or atome - free boolean algebra. I shall mention some other examples. From the dense ordering we can easly pass to the betweenness relation B(x, y, z) on graphs.

If a graph has a circle, then all points of the circle are equivalent with respect to tree places betweenness relation. Then we can introduce 4 places betweenness Be(x, y, z, v). If the graph has finite number of nodes, then its B (or Be) theory is categorical in $alef_0$ and decidable [2]. If the graph has infinite number of nodes, then its theory may be undecidable [1].

Studying models we say that $\mathfrak{M} = \langle M, R \rangle$ is categorical iff the theory of $\mathfrak{M}$ (the set of true sentences in $\mathfrak{M}$) is categorical. Hence (because of its completeness) a theory is categorical iff it is the theory of a categorical model.

Starting from the Ryll-Nardzewski [4] main theorem one can characterize models categorical in $alef_0$ as uniform by decomposition in the following sense:

<u>Definition.</u> $\mathfrak{M}$ is n - parametrically uniform by decomposition if and only if for each n - tuple of parameters $a_1, \dots, a_n \in M$ there is a finite decomposition of M :

$$M = X_1 \cup \dots \cup X_k$$

such that every X_i $(1 \leqslant i \leqslant k)$ is uniform in the logical sense with respect to the parameters $a_1, \dots, a_n$: this means that for every two elements $x, x' \in X_i$ and for every formula Φ with $n + 1$ variables the following equivalence holds:

$$(1) \qquad \mathfrak{M} \models \Phi[x, a_1, \ldots, a_n] \Longleftrightarrow \mathfrak{M} \models \Phi[x', a_1, \ldots, a_n]$$

($\mathfrak{M} \models$ means satisfaction in $\mathfrak{M}$).

<u>Theorem 1.</u> $\mathfrak{M}$ is categorical in alef$_0$ if and only if for every n $\mathfrak{M}$ is n parametrically uniform by decomposition.

Proof in [2].

This theorem involves the possibility of description of models categorical in alef$_0$ using spectrum - trees.

<u>Spectrum - tree.</u> With a model $\mathfrak{M}$ we associate spectrum - trees $T = \langle T, F \rangle$ satisfying the following conditions:

1. The elements of the tree T (the nodes) are subsets of $M : X \in T \longrightarrow X \subset M$
2. F is the function mapping T into T (F (X) is the next node to the root (= top)).

3. The union of the counterimage of every element X is equal to M :

$$\bigcup \left\{ Y : X = F(Y) \right\} = M$$

4. The top element is the empty set :

for every $X \in T$, there is $n \in N$ such that : $F^n(X) = \emptyset$

5. Every $X \in T$ is uniform with repect to parameters $a_1, \dots, a_n$ such that $a_i \in F^i(X)$ and n is the smallest number such that $F^{n+1}(X) = \emptyset$.

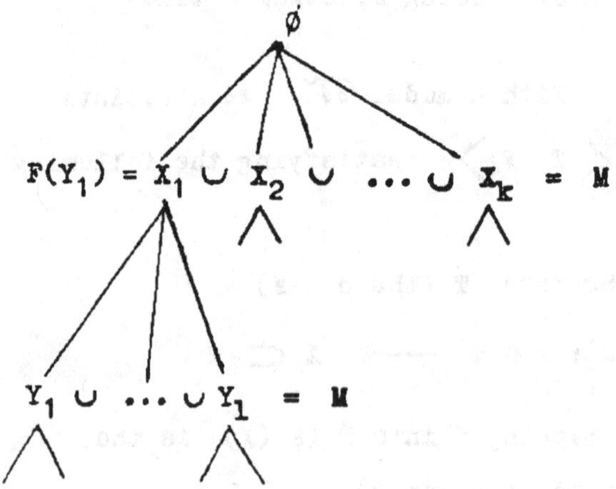

Theorem 2. $\mathfrak{M}$ is categorical in alef$_0$ if and only if there exists an associated finitary spectrum tree, (the counterimage $F^{-1}(X)$, for every $X \in T$ is finite).

Proof. Directly from Theorem 1.

Corollary. In the smallest spectrum tree for $\mathfrak{M}$ categorical in alef$_0$ every element $X \in T$ is definable.

This follows from the proof of Theorem 1.
The smallest spectrum tree is such that the cardinal number of $F^{-1}(X)$ is the smallest one.

2. Decision procedure

The study of spectrum - trees is closely related with the problem of decidability. First from the spectrum tree we can pic up spectrum representatives tree. If $\langle T,F \rangle$ is spectrum tree, then $\langle T', F' \rangle$ is spectrum representatives tree if and only if : from every non empty $X \in T$ we pic up one arbitrary element $a(X)$ and :

$$T' = \left\{ a(X) : X \in T \right\} \cup \left\{ \emptyset \right\}$$

$$F'(a(X)) = \begin{cases} a(F(X)) & \text{when } F(X) \neq \emptyset, \\ \emptyset & \text{when } F(X) = \emptyset \end{cases}$$

Hence $T' \subset M$ and spectrum representatives tree is isomorphic with spectrum tree.

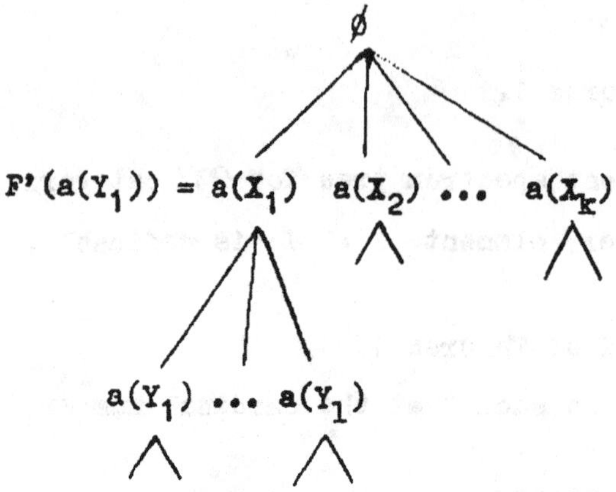

For spectrum representatives tree the following lemma holds :

<u>Lemma 3.</u> If $a_1, \ldots, a_n \in T'$ and $a_{i+1} = F'(a_i)$ for $1 \leqslant i < n$, and $F'(a_n) = \emptyset$, then :

(2) $$\mathfrak{M} \models (x)\ \Phi[x, a_1, \ldots, a_n]$$ if and only if

(3) $$\mathfrak{M} \models \Phi[b_1, a_1, \ldots, a_n] \wedge \ldots \wedge \Phi[b_l, a_1, \ldots, a_n],$$

where $\{b_1, \ldots, b_l\} = F'^{-1}(a_1)$,

for every formula Φ with $n + 1$ free variables,

and the same for existencial quantifier and disjunction respectively.

Proof follows directly from the definition of spectrum - tree and the lemma is generally valid for every model if we allow the conjunction in (3) to be infinitely long when $F'^{-1}(a_1)$ is infinite.

Resting on Theorem 2 we obtain immediately :

Lemma 4. If $\mathfrak{M}$ is categorical in alef_0 , then the conjunction (3) in lemma 3 is finite.

Theorem 5. (On the elimination of quantifiers) If $\mathfrak{M}$ is categorical in alef_0, then for every sentence Ψ there exists a formula Φ without quantifiers (with suitable numer n of free variables) and there exists a finite sequence $a_1, \ldots, a_n \in M$ of elements of T', such that :

(4) $\mathfrak{M} \models \Psi$ iff $\mathfrak{M} \models \Phi[a_1, \ldots, a_n]$

Proof. Applying lemmas 3 and 4 we can replace every general quantifier by a finite conjunction and every existential quantifier by a finite disjunction.

For given model $\mathfrak{M}$ the cardinal number of elements $a_1, \ldots, a_n$ depends on the number of quantifiers in Ψ .

If Ψ contains k quantifiers then the sequence $a_1, \ldots, a_n$ consists of the elements of the first k levels of the tree T'.

For every theory categorical in $alef_0$ which is known, it is easy to describe model with recursive spectrum representatives tree T' in which the primitive relation R on the elements of T' is recursive. Hence from Theorem 5 we get decidability in any known case. This procedure seems to be much simpler then the procedure based on the Vaught general theorem.

3. Conjecture

By conjecture J mean here the following hypothesis :

Conjecture 1. Every theory with finite number of primitive notions categorical in $alef_0$ is decidable.

Accordingly to the theorem 1 this is equivalent to the following one :

Conjecture 2. For every undecidable theory there exists a model $\mathfrak{M}$ and a finite number of elements $a_1, \ldots, a_n \in M$ such that there is infinite sequence of subsets $\{X_i\}$ of the set M which are definable in $\mathfrak{M}$ by some formulas with parameters $a_1, \ldots, a_n$.

The argument for the conjecture: the counter examples fail. Even one can not find a theory categorical in alef$_0$ for which one realy need two quantifiers to pass from n variables predicates to n + 1 variables predicates for infinitely many n. (For boolean atom - free algebra one quantifier is needed). If the number of quantifiers is recursively limited then we can easly estimate the number of all n - variables predicates as a recursive function of n. Hence the tree T' can be recursive.

The problem may be reduced to two-places relation accordingly to the following:

<u>Theorem 6.</u> For every theory T categorical in alef$_0$ with finite number of primitive notions there exists a theory T' such that :

1. T' has two primitive notions : identity and the second one : $x \leqslant y$.

2. The relation $\leqslant$ is assumed in T' to be partial ordering.

3. T' is categorical in alef$_0$

4. T is interpretable in T'.

Proof. Let $\mathfrak{M} = \langle M, p_1, \ldots, p_n \rangle$ be a denumerable model for T. We extend the model $\mathfrak{M}$ to the model $\mathfrak{A} = \langle A, \leqslant \rangle$ by adding new elements in the way used by Rabin and Scott in [3] for proving undecidability. Suppose for simplicity that T has one primitive notion denoting 4 places relation P(x, y, z, u). To every 4-tuple x, y, z, u such that P(x, y, z, u) we add 7 new elements ordered by the relation $\leqslant$ in the following way :

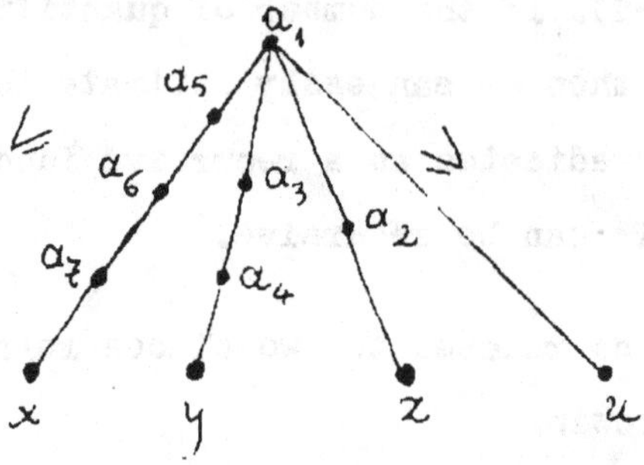

a_1 is maximal, x, y, z, u are minimals. T' is the set of sentences true in $\mathfrak{A}$.

$\leqslant$ is partial ordering. M and P are definable :

(5) $x \in M \Longleftrightarrow (v)(v \leqslant x \longrightarrow v = x)$

(6) $P(x,y,z,u) \Longleftrightarrow x,y,z,u \in M \wedge (E\ a_1, \ldots, a_7)$

$(a_1 > u \wedge a_1 > a_2 > z \wedge$

$a_1 > a_3 > a_4 > y \wedge a_1 > a_5 > a_6 > a_7 > x)$

Hence T is interpretable in T'. Now suppose $\mathfrak{B} = \langle B, \leqslant' \rangle$

be another denumerable model for T'. Define in $\mathfrak{B}$ the substructure $\mathfrak{M}' = \langle M', p_1', \ldots, p_n' \rangle$ consisting of elements satisfying (5) with $\leqslant'$ instead of $\leqslant$, and relations p' are defined in $\mathfrak{M}'$ by the same formulas (say (6)) which are used to define p_i in $\mathfrak{A}$ (with $\leqslant'$ instead of $\leqslant$).

Because of the interpretability of T in T' the structure $\mathfrak{M}'$ is a model for T and hence $\mathfrak{M}'$ is isomorphic with $\mathfrak{M}$. Then this isomorphism can be enlarged to the isomorphism between $\mathfrak{A}$ and $\mathfrak{B}$. Thus T' is categorical in $alef_o$.

The conjecture concerns only the categoricity in $alef_o$. For greater powers there is a counterexample :

Theorem 7. (L.Pacholski) There is a theory of one (one - to - one) function which is udecidable and categorical in any power greater than $alef_o$.

Proof. The primitive notions are: identity and F. For F we assume the following axioms :

1. $(x)(E\,y)\ \ x = F(y)$
2. $F(x) \neq x$
3. $F(x) = F(y) \longrightarrow x = y$

Z is a non recursive set of prime numbers :

4. if $n \in Z$, then $\ulcorner (Ex)\, F^n(x) = x \urcorner$ is an axiom;

5. if $n \notin Z$, then $\ulcorner (x)\, F^n(x) \neq x \urcorner$ is an axiom;

6. if $n = p_1 \cdot \ldots \cdot p_k$ where $p_1, \ldots, p_k$ are primes, then the following formula :

$$\ulcorner (x) \left[F^n(x) = x \longrightarrow (F^{p_1}(x) = x \vee \ldots \vee F^{p_k}(x) = x) \right] \urcorner$$

is an axiom;

7. if p is prime, then the following formula :

$$(F^p(x) = x \wedge F^p(y) = y) \longrightarrow (y = F(x) \vee y = F^2(x) \vee$$

$$\vee \ldots \vee y = F^p(x))$$

is an axiom.

This set of axioms characterizes the function F completely. Every model of the theory consists of a countable set of cercles

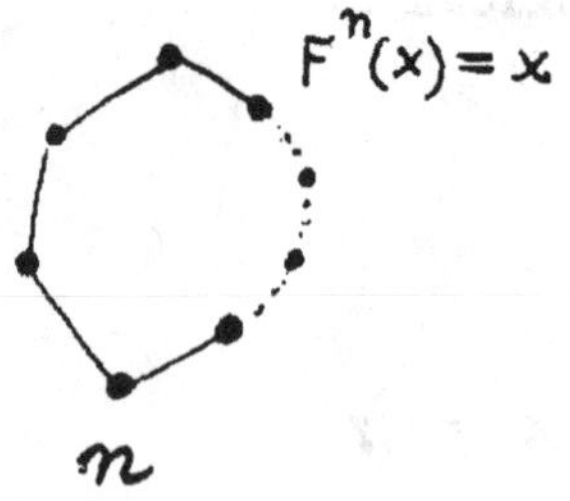

having n - elements for $n \in Z$, and of an arbitrary set of chains

each of the type $\omega^* + \omega$. Hence every model of power $\underset{\sim}{m} >$ alef_0 contains $\underset{\sim}{m}$ chains and the same set of finite circles. Thus every two models of power $\underset{\sim}{m}$ are isomorphic.

References :

[1] H.C.Doets - The relation of succession P on spaces homeomorphic to the cárcle, the open, half - open and closed line [Scriptie], Mathematisch Instituut Universiteit van Amsterdam, November 1966. (Mimeographed).

[2] A.Grzegorczyk - Logical Uniformity by Decomposition and Categoricity in $\aleph_0$. Bulletin de l'Académie Polonaise des Sciences, Série des Sc. math., astr., et phys. vol XVI, No 9, 1968 p.687 - 692.

[3] M.O.Rabin - A simple method for undecidability proofs and same applications. Logic Methodology and Philosophy of Science. Proceedings of the 1964 International Congress. North Holland Publ.Company. p. 58 - 68.

[4] C.Ryll-Nardzewski - On the categoricity in power. Bull.Acad.Polon.Sci., Ser.Sci.math., astr. et phys., 7(1959) p.545 - 548.

[5] R.L.Vaught - Applications of the generalized Skolem-Löwenheim theorem to problems of completeness and decidability. Bull.Amer.Math. Soc.(1953),p.396-397.

On the long-range prospects of automatic theorem-proving

Hao Wang

There is a false contrast between the algorithmic and the heuristic approaches. Every program has to embody some algorithm and for serious advances, partial strategies or heuristic methods are indispensible. Hence, no serious program could avoid either component. Perhaps the contrast is more between anthropomorphic and logicist, as typified by the general problem solver on the one hand and elaborate refinements of the Herbrand theorem on the other. This polarization appears to me to be undersirable and to represent what I would call the reductionist symptom.

Typically the reductionist is struck by the power or beauty of certain modes to proceed and wish to build up everything on them. The two extremes seem to share, in practice if not in theory, this reductionist preoccupation. In my opinion, there should be more reflective examination of the data, viz. the existing mathematical proofs and methods of proof. It is true that what is natural for man need not be natural or convenient for machine. Hence, it will not be fruitful to attempt to imitate man slavishly. Nevertheless, the existing body of mathematics contains a great wealth of material and constitutes the major source of our understanding of mathematical reasoning. The reasonable course would be to distill from this great reservoir whatever is mechanizable. In other words, we should strive for an interplay between reduction and reflection which, for lack of a better name, may be called the dialectic method.

In a previous survey, ([8], 1965), I have set forth a few vague suggestions which are buried in the examples. I should like now to list these suggestions explicitly and use them to make a few remarks on the current scene. (1) It is recommended that powerful methods with restricted ranges of application be explored. (2) Crude strategies are sketched for selecting lemmas in proving theorems of number theory. (3) An example in the predicate calculus is given to illustrate possibilities of directly exploiting special properties of $\equiv$ and local quantifiers (to reduce $\exists x(Fx \wedge x \equiv y)$ to Fy). (4) The need for an adequate treatment of equality is emphasized for both proof procedures and decision procedures in the predicate calculus.

With regard to (4), there have been several proposals during the last few years for adjoining equality to proof procedures of the predicate calculus. In connection with decision procedures, it has turned out that there is a major open theoretical problem, viz. no proof exists in the literature for the belief that there is a decision procedure for the Gödel case with equality. More exactly, the belief is that there is a decision procedure for satisfiability for the class of prenex formulas with equality whose prefix is $\forall x_1 \ldots \forall x_m \exists y_1 \exists y_2 \forall z_1 \ldots \forall z_n$, and, more, that any formula in the class, if satisfiable at all, has a finite model.

With regard to (2), there have been work to carry out the examples from number theory on computers, but only in a weakened form. No strategies are included to select lemmas. Rather, the lemmas are taken as given and a conditional theorem to the effect that the theorem follows from the lemmas is proved as a theorem of the predicate calculus. It is clear that this is not making use of special properties of particular branches in mathematics but rather continuing to ''logicize mathematics''.

In connection with (3), the second proof of ExQ1 ([8], p.55) is intended to give examples of mechanizable special strategies which are suggested by human deductions. The following features are present in the example. (a) Substitute given constants for variables to get stronger conclusions. (b) To eliminate local quantifiers when possible, i.e. strive to introduce a condition $x = y$ to yield $\exists x(x = y \wedge Fx)$ or $\forall x(x = y \supset Fx)$ in order to reduce the quantified expression to Fy. (c) Substitute equivalences freely (if $A_1 \equiv A_2, \ldots, A_{n-1} \equiv A_n$, then A_i can be substituted for A_j). (d) Apply implication chains: $A_1 \supset A_n$ if $A_1 \supset A_2, \ldots, A_{n-1} \supset A_n$. The features (a) and

(d) can be incorporated into Herbrand type proofs fairly directly. But features (b) and (c), though mechanizable and familiar, seem to be destroyed when the problem is transformed into a normal form suitable for obtaining proofs of the Herbrand type. It is thought that by studying examples of human proofs, one may come up with a fair number of useful special strategies such as (b) and (c).

In connection with (1), we may mention the use of least counterexamples in number theory and strategies like (b) and (c) above. In general, it seems desirable to consider directly, besides Skolem functions obtained from dropping quantifiers, also descriptive functions with predetermined meaning such as addition and multiplication in number theory, pair and power set in set theory. It seems desirable to be miserly in the use of quantifiers. In dealing with set theory, it seems desirable to view every axiom of relative existence

$$\exists y \forall x (x \epsilon y \equiv Fxu \ldots v)$$

as defining a function $f_F(u, \ldots, v) = \hat{x}Fxu \ldots v$. In this way, we may operate with constants (such as 0 and ω), functions, and extensionality in form:

$$A \equiv B \supset f_A = f_B .$$

If one reviews the literature on automatic demonstration during the last few years, one gets the impression that the whole field consists of variations on Herbrand's theorem. Often a slight modification is given with full details in a somewhat new dress, accompanied by an elaborate completeness proof. Alternative procedures are offered for alternative advantages. It is hard either to compare the relative efficiency or to accumulate different advantages into one procedure. Hence, some people are looking for a theoretical criterion of relative efficiency. In my opinion, the excessive emphasis on mathematical rigour (completeness proofs, etc.) and purity (theory of efficiency) is a sort of misplaced exactness. A more serious concern is that I do not see how continuing in the same direction, i.e., without thinking more about actual mathematical practice, could lead to major advances.

In the direction of formalization, there are two major successes in modern logic. First, the fairly well established conclusion that all of mathematics is reducible to axiomatic set theory and that, if one takes enough trouble, mathematical proofs can be reproduced in this system completely formally in the sense of mechanical checkability. Second, the results of Skolem and Herbrand according to which we can, by construing mathematical theorems as conditional theorems (viz. that the axioms imply the theorem) in the predicate calculus, search for each mathematical proof in a mechanical (in principle) way to determine whether a related Herbrand expansion contains a contradiction. Impressive as these results are, and encouraging as they are for the project of mechanizing mathematical arguments, they are only theoretical results which do not establish the strong conclusion that mathematical reasoning (or even a major part of it) is mechanical in nature.

What is exciting in the unestablished strong conclusion is that we are facing an altogether new kind of problem which crys out for a totally new discipline and which has wide implications on the perennial problem about mind and machine. We are invited to deal with mathematical activity in a systematic way. Even though what is demanded is not mechanical simulation, the task requires a close examination of how mathematics is done in order to determine how informal methods can be replaced by mechanizable procedures and how the speed of computers can be employed to compensate for its inflexibility. The field is wide open, and like all good things, it is not easy. But one does expect and look for pleasant surprises in this requirement of a novel combination of psychology, logic, mathematics and technology.

It is highly likely that there are different levels of mathematical activity which can be measured by the ease of mechanization. For example, Euler told of how his theorems were often first discovered by empirical and formalistic experimentations. While these experimentations are probably easy to mechanize, the steps of deciding what experimentations to make and of finding afterwards the correct statement and proof of the theorems suggested, are of a higher level and much harder to mechanize. Ramanujan is reported to have commented on the taxicab number 1724 that it is the smallest number expressible as a sum of two cubes in two different ways. The memory and powers of calculation exemplified in this anecdote are probably not hard for a computer, but it would be less easy to have a computer prove most of his theorems. One suspects, however, it would be easier for a computer to prove his theorems than many of the more famous theorems in number theory which are more ''conceptual'' and further removed from calculations. Axiomatic set theory has in more recent years become much more mathematical, and one gets the impression that long formal proofs of relatively simple results are much easier to discover mechanically than advanced neat proofs which can be communicated succinctly between experts.

On the highest level, Poincaré compares Weierstrass and Riemann. Riemann is typically intuitive while Weierstrass is typically logical. In this case, it is natural to believe that it is easier to reach results of Weierstrass mechanically. Hadamard contrasts his impression of the great works of Poincaré and Hermite and states that he finds Hermite's discoveries more mysterious ([4], p. 110). By stretching greatly one's imagination, one might wish to claim that Hadamard would have found it easier to design a program to discover Poincaré's results than to get one for Hermites.

G. Wallas (Art of Thought, 1926, pp.79-107) suggests that there are four stages in the process of bringing about a single achievement of thought: (1) preparation, (2) incubation, (3) illumination, (4) verification. This fits in well with Poincaré lecture on mathematical discoveries (Science and Method). Hadamard ([4]) and Littlewood ([5]) discuss these four stages at great length. The first and the last stages are done consciously. The preparation stage contains two parts: the long-range education of the individual, and the immediate task of learning and digesting what is known about the problem under study. The verification stage consisting of making vague ideas precise and filling in gaps (in particular, carrying out calculations). To mechanize these stages appear formidable enough, but incubation leading to illumination would seem in principle a different kind of process from the operation of existing computers. Since incubation implies an element of rest (an abstention from conscious thought on the initial problem), we may perhaps claim that the importance of this stage comes from a weakness on the part of man, and that machines do not need the period of rest or abstention.

To come back to the current scene, I venture to make some general comments on a few specific aspects. It is appealing to think of an interaction between man and machine, so that computers may become research assistants. In fact, an example of man-machine programs has been written by Guard and others ([3]). It seems that human interventions would be able to improve more substantially the end results if we move from Herbrand proofs to programs with more varied data and strategies.

Practical applications of computers are mainly concerned with repetitions of simple steps rather than individualized long sequences of simple steps such as mathematical proofs. It is natural to think of applying mechanical inference to cases where a lot of short deductions are made. For example, it has been suggested that we can retrieve simple consequences of stored information on individual persons (e.g., Darlington, [2]).

Suggestions have been made to extend automatic demonstration to higher-order logic. It is, however, not clear to me why this could be considered more promising than looking directly at, say, number theory or axiomatic set theory which, in my opinion, is more suggestive and closer to real life. Usable examples in set theory can be found in [8], 1967.

The central idea of automatic demonstration during the last few years appears to be the observation that in order to derive a contradiction from the Herbrand expansion of a formula, it is sufficient to examine mechanically all possible substitutions to obtain potential contradictions. It was noted by Prawitz ([6]) that we can devise an algorithm to a decide whether, given a conjunction C of finitely many clauses and a recursive set of terms, there exists a substitution of terms for variables in C such that the result contains a contradiction. Moreover, given any partition of all terms in C into equivalence classes, there is a least or most general substitution, if there is any, that yields the partition: α is the least if for any β yielding the same partition, we can find γ, $\gamma\beta C \equiv \alpha C$. This idea was applied independently by Robinson ([7]) and Aanderaa ([1]) to introduce what is called resolution (by Robinson) or generalized cut (by Aanderaa). Various generalizations and refinements of the ''resolution method'' have been proposed.

Elsewhere, I have stressed the advantage of ''miniscope'' form. In this way, the Skolem functions resulting from existential quantifiers in general get fewer argument variables than in the usual prenex form (compare reference number 10 of [8]). This is adopted in Aanderaa's algorithm. Aanderaa also uses ''generalized contraction'' and a priority function to govern the order in which different clauses are ''confronted'' to yield generalized cuts. Unfortunately, I am not able to follow all his intricate steps to give a reasonable sketch of his detailed methods.

References

[1] S. Aanderaa, A deterministic proof procedure (manuscript of a term paper), 61 pp., Harvard, May, 1964.

[2] J.L. Darlington, ''Theorem proving and information retrieval'', Machine intelligence, vol.4 (1969), Edinburgh.

[3] J.R. Guard, J.H. Bennett, W.B. Easton, L.G. Settle, ''CRT-aided semi-automated mathematics'', AFCRL-67-0167, 1967.

[4] J. Hadamard, Psychology of invention in the mathematical field, Princeton, 1945.

[5] J.E. Littlewood, ''The mathematician's art of work'', The Rockefeller University Review, September-October, 1967, New York.

[6] D. Prawitz, ''An improved proof procedure'', Theoria, vol.26 (1960), pp.102-139.

[7] J.A. Robinson, ''A machine-oriented logic based on the resolution principle'', J.ACM, vol.12 (1965), pp.23-41.

[8] H. Wang, ''Formalization and automatic theorem-proving'', Proc.IFIP Congress, 1965, vol.1, pp.51-58; ''Examples in set theory'', Z.f.Logik u.Grundl.d.Math., vol.13 (1967), pp.175-188, 241-250.

THE CASE FOR USING EQUALITY AXIOMS IN AUTOMATIC DEMONSTRATION

Robert Kowalski

Introduction.

The use of equality axioms in resolution refutation systems has seemed to be particularly inefficient. In order to remedy this difficulty several modifications of the resolution method have been proposed ([4] , [13] , [15] , [17] and [21] and more recently [2] and [10]). Of these the paramodulation strategy of [15] seems to be particularly simple and efficient. The method for dealing with equality investigated in this paper consists of using equality axioms and of applying the version of hyper-resolution proposed in [5] . The hyper-resolution and paramodulation methods are compared and a simple interpretation of the former is found in a subsystem of the latter, providing a straightforward proof for the completeness of this subsystem of paramodulation. Several proposals are put forward for modifying the hyper-resolution method and these modifications are seen to induce corresponding modifications of the paramodulation strategy.

The method of this paper need not be confined to equality and can be applied to the special treatment of more general sets of axioms.

Preliminaries.

If L is a literal then $|L|$ denotes the atom A such that $L = A$ or $L = \bar{A}$. An expression (literal, clause, set of clauses) is a ground expression if it contains no variables. Constants are function symbols with no arguments. A set of expressions E is unifiable with unifier σ if $E\sigma$ is a singleton. If E is unifiable then there is a substitution θ , called a most general unifier (m.g.u.) of E, such that θ unifies E and for any unifier σ of E, $\sigma = \theta\lambda$ for some λ . Similarly a family of sets of expressions $\mathcal{E}$ is simultaneously unifiable with simultaneous unifier σ if $E\sigma$ is a singleton for each $E \in \mathcal{E}$. If $\mathcal{E}$ is simultaneously unifiable then there exists a simultaneous unifier θ of $\mathcal{E}$ such that for any simultaneous unifier σ of $\mathcal{E}$, $\sigma = \theta\lambda$ for some λ ; θ is called a most general simultaneous unifier (m.g.s.u.) of $\mathcal{E}$.

A set of ground clauses $\mathcal{C} = \{A_1,\ldots,A_n,B\}$ where, for $1 \leq i \leq n$, $A_i = \{L_i\} \dot{\cup} A_{0i}$ and $B = \{\bar{L}_1,\ldots,\bar{L}_n\} \dot{\cup} B_0$ is a clash ($\dot{\cup}$ denotes disjoint union as in Andrews' [1]). The resolvent of $\mathcal{C}$ is the clause $C = A_{01} \cup \ldots \cup A_{0n} \cup B_0$. The clauses in $\mathcal{C}$ are the parents of C; $A_1,\ldots, A_n$ are the satellites and B the nucleus of $\mathcal{C}$. The literals $L_1,\ldots,L_n$ and their complements $\bar{L}_1,\ldots\bar{L}_n$ are said to be literals resolved upon in $\mathcal{C}$. $\mathcal{C}$ is restricted if $L \not\in C$ when L is resolved upon in $\mathcal{C}$. If $\mathcal{C}$ contains only two clauses then $\mathcal{C}$ and C are said to be binary. A clause is positive if all its literals are positive (i.e. unnegated atoms). The resolvent of a restricted clash is called a <u>hyper-resolvent</u> if all its satellite parents are positive and if all the negative literals in its nucleus parent are resolved upon in the clash. A <u>P_1-resolvent</u> is a binary resolvent with one positive parent. Every hyper-resolvent can be obtained by a sequence of P_1-resolutions [11].

To define the general notion of clash (called latent clash in [12] and [18]) we first define the notion of factor of a clause. The definition below, introduced in [5], differs both from that of Wos and Robinson [20] and from that more recently introduced by J.A. Robinson in [14]. Given a clause A and a unifiable subset $E \subseteq A$ with m.g.u. θ then the clause $A\theta$ together with its <u>distinguished literal</u> $E\theta$ is called a <u>satellite factor</u> of A. Given a clause B and a unifiable family $\mathcal{E}$ of subsets of B with m.g.s.u. θ then the clause $B\theta$ together with the set of its <u>distinguished literals</u> $\mathcal{E}\theta$ is called a <u>nucleus factor</u> of B. In case $\mathcal{E}$ contains only one set of literals then the corresponding nucleus factor of B is also a satellite factor. Conversely any satellite factor can be regarded as a nucleus factor. We write the distinguished literals of a factor as its first literals. A nucleus factor is <u>complete</u> if the set of its distinguished literals coincides with the set of its negative literals.

For the general case we define clash resolution for sets of factors and we insist that all and only distinguished literals be resolved upon. Given n satellite factors of the form $A_1 = \{L_1\} \dot{\cup} A_{01},\dots,A_n = \{L_n\} \dot{\cup} A_{0n}$ and nucleus factor $B = \{\overline{K}_1,\dots,\overline{K}_n\} \dot{\cup} B_0$ (where none of the factors $A_1,\dots,A_n,B$ share common variables), the set $\mathcal{C} = \{A_1,\dots,A_n,B\}$ is a <u>clash</u> if the family $\mathcal{E} = \{\{L_1,K_1\},\dots,\{L_n,K_n\}\}$ is unifiable. If $\mathcal{C}$ is unifiable with m.g.s.u. θ, then $C = (A_{01} \cup \dots \cup A_{0n} \cup B_0)\theta$ is the <u>resolvent</u> of $\mathcal{C}$. $\mathcal{C}$ is <u>restricted</u> if $L\theta \notin C$ when L is resolved upon in $\mathcal{C}$ and θ is the m.g.s.u. of $\mathcal{C}$. Satellites, nucleus, hyper-resolvent, etc. are defined as for ground clauses. Notice that the nucleus parent of a hyper-resolvent is always complete. It is easy to show, as in [5], that these notions of factoring and clash resolution improve the usual notions by restricting the generation of redundant inferences and of repeated calculation of m.g.s.u.s.

An arbitrary set of factors $\mathcal{C}$ is a clash if some set $\mathcal{C}'$ of variants of factors in $\mathcal{C}$ is a clash, i.e. $\mathcal{C}'$ is $\mathcal{C}\sigma$ where $A\sigma$ is a variant of A for each $A \in \mathcal{C}$ and where no two factors in $\mathcal{C}'$ contain common variables. The resolvent of $\mathcal{C}$ is the resolvent of $\mathcal{C}'$. Similarly an arbitrary set of clauses $\mathcal{C}$ is a clash if some set of factors $\mathcal{C}'$ is a clash where $\mathcal{C}'$ contains exactly one factor for each clause in $\mathcal{C}$ and where each factor in $\mathcal{C}'$ is a factor of some clause in $\mathcal{C}$. Again the resolvent of $\mathcal{C}$ is the resolvent of $\mathcal{C}'$.

Let S be a set of factors, define $\mathcal{R}(S)$ as the union of S and of the set of satellite factors of binary resolvents whose parents belong to S. Define $\mathcal{H}(S)$ similarly as the union of S and of the set of satellite factors of hyper-resolvents whose parents belong to S. For unfactored clauses S let $\mathcal{R}^0(S)$ be the set of all satellite factors of clauses in S, let $\mathcal{H}^0(S)$ be the set of all satellite factors of positive clauses in S and of all complete nucleus factors of non-positive clauses in S. Given any operation $\mathcal{O}$ (such as $\mathcal{R}$ or $\mathcal{H}$) from sets of factors to sets of factors and operation $\mathcal{O}^0$ (such as $\mathcal{R}^0$ or $\mathcal{H}^0$) from sets of unfactored clauses to sets of factors define $\mathcal{O}^{n+1}(S) = \mathcal{O}(\mathcal{O}^n(S))$ for $n \geq 0$. The completeness theorems for binary resolution and for hyper-resolution state that given an unsatisfiable set of clauses S_0 then $\square \in \mathcal{R}^n(S_0)$ and $\square \in \mathcal{H}^m(S_0)$ for some $n \geq 0$ and $m \geq 0$.

Let $\mathcal{O}$ be a derivation operation such as $\mathcal{R}$ or $\mathcal{H}$ above then given clauses S and $C \in \mathcal{O}^n(S)$ for some n there corresponds to C in a natural way at least one derivation tree T of C from clauses in S. It is convenient not to exhibit in T the operation of factoring or the operation of standardising variables in clauses occurring in clashes. We shall say that C occurs at the <u>root</u> of T and clauses from S at the <u>tips</u> of T. We shall think of T as a partially ordered set of occurrences of clauses with the occurrences at the tips being maximal elements and the occurrence at the root being the least element lying below all others. Given such a tree T and $C \in T$ we call T' the <u>subtree of T rooted in C</u> if T' consists of all of T lying above C and including C at its root. Thus if T is a derivation from clauses in S then any such T' is also a derivation from clauses in S.

<u>Equality Axioms and the Hyper-Resolution Method.</u>

For the equality symbol $=$ we write $s = t$ instead of $=(s,t)$ and $s \neq t$ for $\overline{s = t}$.

Let S_0 be a set of clauses. Let $E = E_1 \cup E_2 \cup E_3$ where

$E_1 = \{ x = x \}$,

$E_2 = \{ x_i \neq y_i, f(x_1,\ldots,x_i,\ldots,x_n) = f(x_1,\ldots,y_i,\ldots,x_n) \mid$ for f in the vocabulary of S_0, for $n \geq 1$ and $1 \leq i \leq n \}$,

$E_3 = \{ x_i \neq y_i, \overline{P}(x_1,\ldots,x_i,\ldots,x_n), P(x_1,\ldots,y_i,\ldots,x_n) \mid$ for P the equality symbol and for P in the vocabulary of S_0, $n \geq 1$ and $1 \leq i \leq n \}$.

For simplicity we adopt the convention that $s = t$ is syntactically indistinguishable from $t = s$.

If S_0 has no normal model (i.e. a model in which the equality symbol of S_0 is interpreted as a substitutive identity relation) then $S = S_0 \cup E$ has no model whatsoever. Therefore there exists a hyper-resolution derivation T of $\square$ from S.

The efficiency of obtaining T can be improved in several directions by imposing restrictions on the hyper-resolution method [5]. Among the more important of these is the α-restriction ([5] and [6]). Given a set of clauses S, $\leq$ is an α-ordering for S if $\leq$ is a partial ordering of the set of atoms constructible from the vocabulary of S such that for all substitutions σ,

$$L_1 \leq L_2 \text{ implies } L_1\sigma \leq L_2\sigma.$$

Given S and $\leq$ an α-ordering for S, a satellite factor $C \in \mathcal{H}^n(S)$ for some $n \geq 0$, $C = \{L_1\} \dot{\cup} C_0$, is an <u>$\alpha$-factor</u> if $|L_1| < |L_2|$ for no $L_2 \in C_0$. For all n let $\mathcal{H}_\alpha^n(S)$ be $\mathcal{H}^n(S)$ without satellite factors which are not also α -factors. Then S unsatisfiable implies that $\Box \in \mathcal{H}_\alpha^n(S)$ for some n. (1)

If, for example, $\leq$ orders equality atoms before all others then the α-restriction for $\leq$ implies that we need never generate satellite factors of a clause containing the equality symbol if the clause also contains other predicate letters distinct from the equality symbol.

The derivation T of $\Box$ from $S_0 \cup E$ may be taken to be by hyper-resolution with or without the α-restriction. We shall compare the efficiency of obtaining T to that of obtaining a refutation of S_0 by the Robinson - Wos system of paramodulation and resolution.

<u>Paramodulation and its Completeness.</u>

Given a clause B and a single occurrence of the term t in B we write B(t) to indicate the given occurrence of t in B. For ground level clauses A = $\{t = s\} \dot{\cup} A_0$ and B = B(t) a paramodulant of A and B is the clause C = $B(s/t) \cup A_0$. B(s/t) indicates the result of replacing the distinguished occurrence of t in B(t) by s. When there is no possibility of confusion we shall also indicate the result of this replacement by B(s).

At the general level paramodulation is defined in the context of refutation systems which include a separate rule for factoring. For factors A = $\{t_1 = s\} \dot{\cup} A_0$ and B = $B(t_2)$ where A and B share no variables and where t_1 and t_2 are unifiable with m.g.u. θ, a <u>paramodulant</u> of A and B is the clause C = $A_0\theta \cup B\theta(s\theta/t_2\theta) = (A_0 \cup B(s/t_2))\theta$. We shall refer to the factors A and B respectively as the <u>first and second parents</u> of C. We call the distinguished occurrence of t_2 in B the <u>term paramodulated upon</u>, although in a more precise terminology it would be more cumbersomely referred to as "the distinguished occurrence of the term paramodulated upon". The literals in which the distinguished occurrences of t_1 and t_2 appear are called the literals <u>paramodulated upon.</u> We shall see that these literals may be taken to be precisely the distinguished literals of satellite factors.

An important case of paramodulation occurs when the term t paramodulated upon in a second parent B(t) is primary in B(t). An occurrence of a term t in B is

---

(1) The same theorem holds if $\mathcal{H}$ is replaced by $\mathcal{R}$, [5]. A weaker theorem holds for set of support.

<u>primary</u> if that occurrence is as an argument of some literal in B [17]. Thus, for example, if $B = \{f(c) \neq c\}$ then f(c) and the second occurrence of c are primary in B but the first occurrence of c is not. We shall say that an application of paramodulation is <u>primary</u> if the term paramodulated upon in the second parent is primary.

We extend the definition of paramodulation to arbitrary factors and to unfactored clauses just as in the case of clashes.

Given clauses S_0 let, E_4 be the set consisting of the clauses $f(x_1,\ldots,x_n) = f(x_1,\ldots,x_n)$ for every function symbol f occurring in the Herbrand universe $H(S_0)$, $n \geq 0$. Let $\mathcal{U}^0(S_0 \cup E_4)$ be the set of satellite factors of clauses in $S_0 \cup E_4$. Let $\mathcal{U}(S)$, for any set of factors S be the union of S and of the set of satellite factors of binary resolvents and of paramodulants whose parents belong to S. The following is a version of the completeness theorem reported in [15].

<u>Theorem 1.</u> Suppose that the set of clauses S_0 has no normal model. Then for some $n \geq 0$, $\square \in \mathcal{U}^n(S_0 \cup E_4)$.

<u>Comparison of the Paramodulation and Hyper-Resolution Methods.</u>

Let S_0 have no normal model and let T be a hyper-resolution derivation of $\square$ from $S = S_0 \cup E$. The clauses in $E_2 \cup E_3$ occur in T only as nuclei of clashes. Moreover we may insist that no clause in E_3 occurring in T be factored with its two negative literals unified; such a factor would be of the form $\{s \neq t, t = s\}$ and would therefore be eliminable as a tautology. Thus to each clause C in E there corresponds exactly one factor. In case C is in E_2 or E_3 then the distinguished literals in the corresponding factor of C are just the negative literals in C. If $C = \{x = x\}$ then $x = x$ is the distinguished literal of the corresponding factor of C. We may identify without confusion the set of clauses E with the set of corresponding factors. The basis for comparison of the hyper-resolution and paramodulation methods rests upon the following two observations (similar observations have been made independently by Chang in [3]):

(1) Every hyper-resolvent in T with nucleus parent in E_3 is a primary paramodulant of its two satellite parents.

(2) Every hyper-resolvent in T with nucleus parent B in E_2 is a paramodulant of its one satellite parent and of an appropriate factor B^* in E_4 (if $B = \{x_i \neq y_i, f(x_1,\ldots,x_i,\ldots,x_n) = f(x_1,\ldots,y_i,\ldots,x_n)\}$ then $B^* = \{f(x_1,\ldots,x_n) = f(x_1,\ldots,x_n)\}$).

To verify (1) suppose that $A = \{s = t\} \dot{\cup} A_0$ and $B = \{P(s_1,\ldots,s_i,\ldots,s_n)\} \dot{\cup} B_0$ are satellite factors of a clash in T having nucleus in E_3. The hyper-resolvent of this clash is the clause $C = (\{P(s_1,\ldots,t,\ldots,s_n)\} \cup B_0)\theta$ where θ is the m.g.u. of the set of terms $\{s,t\}$. But C is also a primary paramodulant of A and B.

To verify (2) let $A = \{s = t\} \dot{\cup} A_0$ be the satellite factor of a clash in T having nucleus B in E_2. The hyper-resolvent of the clash is the clause

$= \{f(x_1,\dots,s\ ,\dots,x_n) = f(x_1,\dots,t,\dots,x_n)\} \cup A_0$ for the function letter f
nd index i of the nucleus B. But C is also a paramodulant whose first parent is
and second parent is the clause $B^* = \{f(x_1,\dots,x_n) = f(x_1,\dots,x_n)\} \in E_4$.

For application below we note that (1) and (2) above hold for resolvents of rbitrary clashes $\mathfrak{C}$ provided only that the literals resolved upon in the nucleus f $\mathfrak{C}$ coincide with the subset of its negative literals. In other words, (1) nd (2) continue to hold even if $\mathfrak{C}$ is unrestricted and the satellites of $\mathfrak{C}$ are ot positive.

Suppose S_0 has no normal model, then $S = S_0 \cup E$ is unsatisfiable but so is $' = S_0 \cup E_2 \cup E_3 \cup E_4$. Let T' be a hyper-resolution derivation of $\square$ from S'. bservations (1) and (2) hold equally well for T' as for T. We shall transform ' into a paramodulation and binary resolution derivation T" of $\square$ from $S_0 \cup E_4$. elete all tips of T' at which clauses from E_3 occur; replace each clause B rom E_2 occurring at a tip of T' by the appropriate clause B^* from E_4. The esulting tree is a paramodulation and hyper-resolution derivation of $\square$ from $_0 \cup E_4$. If we now decompose each remaining hyper-resolvent into a sequence of $_1$-resolvents (see [5] or [11]) we obtain the desired derivation T". The act that such a derivation T' can always be transformed into a corresponding erivation T" constitutes a proof of Theorem 1. By investigating the structure f T" more closely we see that we have in fact proved the much stronger theorem below.

Let S_0 be a set of clauses and $\leq$ an α-ordering for S_0. Associate with very complete nucleus factor B of a non-positive clause in S_0 a single total rdering of the negative (i.e. distinguished) literals in B. The definitions elow of $\mathcal{U}^0_\alpha$ and $\mathcal{U}_\alpha$ are formulated to guarantee that each P_1-resolvent in " is obtained by decomposing a hyper-resolvent in T' in a unique way. This unique ecomposition is accomplished by resolving on the distinguished literals of non-ositive factors in the order imposed by the original total ordering given to he distinguished literals in complete factors of clauses in S_0. Totally rdering complete nucleus factors eliminates all but one of the n! ways of ecomposing hyper-resolvents whose nucleus parent contains n distinguished iterals.

Let $\mathcal{U}^0_\alpha(S_0)$ be the set consisting of

(1^0) all α-factors of positive clauses in S, and

(2^0) all complete factors of all non-positive clauses in S.

Let $\mathcal{U}_\alpha(S)$, for any set of factors S be the union of S and of the set of onsisting of

(1) all α-factors of paramodulants whose parents are α-factors in S, and

(2) for every P_1-resolvent C both of whose parents are in S and one of which is an α-factor A and the other of which, B, is non-positive and is resolved upon its first distinguished literal

(a) if C is positive then all α-factors of C, and

(b) if C is non-positive then exactly one factor consisting of the clause C; the distinguished literals of C descend from the distinguished literals of B and are totally ordered by the ordering inherited from that of the distinguished literals of B.

In obtaining $C \in \mathcal{U}_\alpha(S)$ we insist that only distinguished literals be paramodulated or resolved upon.

Theorem 2. Suppose that the set of clauses S_0 has no normal model and that $\leq$ is an α-ordering for S_0. Then for some $n \geq 0$, $\Box \in \mathcal{U}_\alpha^n(S_0 \cup E_4)$.

Theorem 2 follows from the fact that we can insist that satellite factors of clashes in T' be α-factors. Further examining the construction of T" from T' we see that Theorem 2 continues to hold withthe following restrictions:

(r1) All applications of paramodulation are primary except when the second parent is a clause $B \in E_4$ in which case the term paramodulated upon is one of the arguments of the function symbol occurring in B.

(r2) If $B \in E_4$ then B is not a first parent of a paramodulant and as parent of a resolvent it may be replaced by the clause $\{x = x\}$.

(r2) follows by constructing the tree T" directly from T instead of from T'.

It should now be fairly clear that the paramodulation method of Theorem 2 incorporating restrictions (r1) and (r2) is, in a sense, isomorphic to the method of hyper-resolution with equality axioms. From this fact it follows that any strategy compatible with the hyper-resolution method translates into a strategy compatible with paramodulation. Clearly deletion of tautologies and deletion of subsuming clauses are two such strategies. Moreover completeness for hyper-resolution of renaming [8] implies the same for paramodulation provided the equality symbol itself is not renamed.

Semantic Trees.

The notion of semantic tree was formulated by Robinson in [13] and investigated in the version described below in [5]. We include a summary of definitions and propositions for the special case of semantic tree used in the proofs of Theorems 3, 4 and 5.

Semantic trees are finitely branching but possibly infinite trees with the root as greatest element lying above all other nodes including any tips which are minimal elements. Thus regard semantic trees as growing downwards rather than upwards as in the case of derivation trees. Given a totally ordered (finite or infinite) set of ground atoms $K = \{A_1, \ldots, A_n, \ldots\}$ where $i<j$ implies that A_i precedes A_j, a binary semantic tree for K is a binary tree T with sets of literals attached to its nodes, such that

(1) $\emptyset$ is attached to the root, and

(2) if $\{A_n\}$ or $\{\overline{A}_n\}$ is attached to the node N then $\{A_{n+1}\}$ and $\{\overline{A}_{n+1}\}$ are attached to the two nodes lying immediately below N.

Given $N \in T$, $\mathfrak{a}_N$ called the <u>assignment at N</u>, is the set consisting of the literals attached to N and to all nodes lying above N in T. A clause C <u>fails</u> at N if for some ground substitution σ the complements of all the literals in $C\sigma$ occur in $\mathfrak{a}_N$. If S is a set of clauses then N is a <u>failure point for S</u> if some $C \in S$ fails at N but no $D \in S$ fails at any node above N. If no $C \in S$ fails at N then N is <u>free for S</u>. T is <u>closed for S</u> if every path beginning at the root and terminating, if at all, only at a tip of T contains a failure point for S. N is an <u>inference node for S</u> if the nodes immediately below N are failure points for S. If S is unsatisfiable and K includes all atoms in some ground unsatisfiable set of instances S' of instances of clauses in S then we say that T is a <u>semantic tree for S</u> (relative to K).

If S is unsatisfiable and T is a semantic tree for S then T is closed for S. If T is closed for a set of clauses S then S is unsatisfiable and moreover T contains at least one inference node N for S. Some clause $C \in \mathcal{R}(\mathcal{R}^0(S))$ fails at N and its parents in S fail at the nodes immediately below N. More generally given any node $N \in T$ free for S then for some n and $C \in \mathcal{R}^n(S)$, C fails at N. Only $\square$ fails at the root of T.

<u>Trivialisation of Inequalities.</u>

Resolving a factor $C = \{s \neq t\} \dot{\cup} C_0$ with $\{x = x\}$ produces the clause $C_0\theta$ where θ is an m.g.u. of the set $\{s, t\}$. We shall call such a resolution the operation of trivialising an inequality [17]. Application of this operation is necessary for both the hyper-resolution and paramodulation methods. Corollary 1 of the more general Theorem 3 below states in effect that Theorem 2 with restrictions (r1) and (r2) continues to hold with the restriction.

(r3) No inequality in a clause C is trivialised unless either C belongs to the original set S_0 or else C is itself the result of trivialising an inequality.

Strictly speaking Theorem 2 with restrictions (r1) - (r3) needs to be modified to allow the trivialisation of distinguished literals other than just the first distinguished literal of non-positive factors.

<u>Theorem 3</u>. Suppose $S = S_0 \dot{\cup} S_1$ is unsatisfiable where S_1 is a satisfiable set of unit clauses. Then the set $S_0 \cup R$ is unsatisfiable where R is the set of resolvents of clashes with nuclei in S_0 and satellites in S_1.

<u>Proof</u>. Assume first that S is a set of ground clauses. Let $S_1 = \{\{L_1\}, \ldots, \{L_n\}\}$. We prove the theorem for this case by showing by induction that for all $k \leq n$:

$U_k = S_0 \cup R_k \cup (S_1 - \{\{L_1\}, \ldots, \{L_k\}\})$ is unsatisfiable where R_k is the set of resolvents of clashes with nuclei in S_0 and satellites in $\{\{L_1\}, \ldots, \{L_k\}\}$.

$U_0 = S$ is unsatisfiable. Suppose that U_k is unsatisfiable for $0 \le k \le n$. Let T be a binary semantic tree for U_k relative to the set of atoms occurring in U_k ordered in such a way that the atom $|L_{k+1}|$ occurs last. T is closed for U_k; we claim that T is also closed for U_{k+1} and that U_{k+1} is therefore unsatisfiable. If $\{L_{k+1}\}$ fails on T at a failure point for U_k then it fails below an inference node N for U_k. The clause failing at the second failure point below N belongs to $S_0 \cup R_k$. The resolvent of these two clauses belongs to R_{k+1} and fails at N. Since $R_k \subseteq R_{k+1}$, it follows that T is closed for U_{k+1} and that U_{k+1} is unsatisfiable. But then $U_n = S_0 \cup R$ is unsatisfiable.

If S is not a set of ground clauses let $S' = S_0' \dot{\cup} S_1'$ be an unsatisfiable set of ground instances of clauses in S where S_0' and S_1' are instances of clauses in S_0 and S_1 respectively. Then $S_0' \cup R'$ is unsatisfiable where R' is the set of resolvents of clashes with nuclei in S_0' and satellites in S_1'. But by the lifting lemma for clashes, $S_0' \cup R'$ is a set of ground instances of $S_0 \cup R$, which is therefore unsatisfiable.

<u>Corollary 1.</u> Suppose $S = S_0 \cup E$ is unsatisfiable. Then $S_0 \cup R \cup E_2 \cup E_3$ is unsatisfiable where R is the set of resolvents of clashes with nuclei in S_0 and with satellites in E_1.

<u>Proof.</u> Take $S_0 \cup E_2 \cup E_3$ above to be the S_0 of Theorem 1. Then $U = S_0 \cup R_0 \cup E_2 \cup E_3$ is unsatisfiable where R_0 is the set of resolvents of clashes with nuclei in $S_0 \cup E_2 \cup E_3$ and satellites E_1. We shall show that resolvents of clashes with nuclei in E_2 or E_3 can be removed from R_0 without affecting the unsatisfiability of U.

Let S' be an unsatisfiable set of ground instances of clauses in S. We may choose S' so that it contains no tautologies and no instances D' of a clause in E_2 where D' is of the form $\{t_i \ne t_i, f(t_1,\ldots,t_n) = f(t_1,\ldots,t_n)\}$, because such an instance of D is subsumed by the instance $\{f(t_1,\ldots,t_n) = f(t_1,\ldots,t_n)\}$ of $\{x = x\}$. We may assume that each resolvent C in R_0 is obtained by lifting a clash whose nucleus B' and satellites belong to S'. But then it is easy to verify that if C is the resolvent of a clash with nucleus B in E_3 then the corresponding instance B' of B in S' is a tautology and C may be eliminated from U. Similarly if $C \in R_0$ is the resolvent of a clash with nucleus B in E_2 then the corresponding instance B' of B in S' is of the form $\{t_i \ne t_i, f(t_1,\ldots,t_n) = f(t_1,\ldots,t_n)\}$ and B may be eliminated from U.

Thus every clause in $R_0 - R$ may be removed from U without affecting its unsatisfiability. It follows therefore that the set of clauses $S_0 \cup R \cup E_2 \cup E_3$ is unsatisfiable.

Both Theorems 3 and 5 below are versions of the throw-away strategies discussed in [9].

Theorem 3 implies that satisfiable sets S_1 of unit clauses may be effectively reprocessed out of a set of clauses $S = S_0 \dot{\cup} S_1$ before attempting to find a efutation of the resulting set S_0*. Our intuition is that such preprocessing s likely to increase the efficiency of obtaining proofs of more difficult theorems. he figure below gives a simple example of two derivations of the same clause. nly the first derivation will be generated if the original set S_0 is pre- rocessed. If the entire set S_0* must be generated before attempting to find a efutation then this method of preprocessing may be inefficient for proving heorems which have a simple proof which can be detected for instance with less ffort than that involved in generating all of S_0* itself. On the other hand ince resolving a clause $A \in S_0$ with a unit clause in S_1 produces a clause containing ewer literals than are contained in A we may expect that this preprocessing rocedure will tend to retain the simplest of those derivations which differ by ermuting occurrences of clauses from S_1 along their branches. Finally even or the case of simpler theorems preprocessing can be made more efficient by imultaneously generating S_0* and generating resolvents from S_0*.

xample.

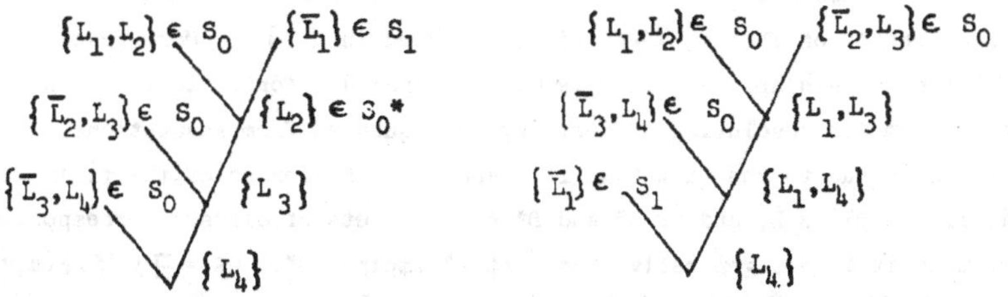

Notice that the redundancy exemplified here can not be removed by implementing singly connectedness [21] and is not removed by hyper-resolution, since both lerivations are by hyper-resolution.

The Subsumption Theorem.

The following theorem, applied in the proof of Theorem 5 below, is of independent interest because it provides partial information about the extent of the leduction completeness of resolution.

Theorem 4. Let S be a non-empty set of clauses and C a non-tautologous clause logically implied by S. Then for some $n \geq 0$ there is a clause $C' \in \mathcal{R}^n(S)$ which subsumes C. Moreover if C is positive then $C' \in \mathcal{H}^m(S)$, for some $m \geq 0$.

Proof. If S implies C then $S \cup \{\overline{C}\}$ is unsatisfiable, where the sentence $\overline{C}$, in prenex normal form, is existential and its matrix is a conjunction of literals. S contains no existential quantifiers. Eliminating existential quantifiers from $S \cup \{\overline{C}\}$ we obtain an unsatisfiable set of clauses $S_0 = S \cup \{\{\overline{L}_1\}, \ldots, \{\overline{L}_n\}\}$ where each $\overline{L}_i$ is a ground literal. Notice that C can be reobtained from $\{L_1, \ldots, L_n\}$ by applying the substitution σ^* which substitutes for each Skolem constant a the variable x in $\overline{C}$ which was replaced by a when eliminating existentially quantified variables.

Let T be a binary semantic tree for S_0 relative to some set of ground atoms K containing all atoms in some ground unsatisfiable set S_0' of instances of clauses in S_0. Let K be ordered in such a way that the atoms $|L_1|, \ldots, |L_n|$ precede all other atoms in K. Let T' be the subtree of T rooted in the node N to which is assigned the set of literals $\{\overline{L}_1, \ldots, \overline{L}_n\}$ (N exists because C is not a tautology). T and T' are both closed for S_0. Moreover, since no clause $\{\overline{L}_i\}$ fails in T', T' is closed for S and some $C' \in \mathcal{R}^n(S)$ fails at N. But then, by the definition of failure, there is a substitution σ such that $C'\sigma \subseteq \{L_1, \ldots, L_n\}$. Let $\sigma' = \sigma\sigma^*$ then $C'\sigma' \subseteq \{L_1, \ldots, L_n\}\sigma^* = C$ and therefore C' subsumes C.

If C is positive we take T to be an M-clash tree for S_0 where M is the set of all negative ground literals which are complements of the atoms occurring in some unsatisfiable set S_0' of ground instances of clauses in S_0(see [5]). Let T' be any subtree of T rooted in a node to which is assigned the set of literals $\{\overline{L}_1, \ldots, \overline{L}_n\}$. Some $C' \in \mathcal{H}^n(S)$, for some $n \geq 0$, fails at the root of T'. It follows that C' subsumes C.

A weaker version of Theorem 4 was first reported in [7]. More recently a more general theorem has appeared in [19]. Theorem 4 unfortunately does not settle the problem for resolution of deriving consequences from assumptions. That this is so is due to the fact that if A and B are sentences of the first-order logic, if A implies B, and if A* and B* are the sets of clauses corresponding to A and B, then it is not generally true that A* implies B*. $A = \exists y \forall x P(x,y)$ and $B = \forall x \exists y P(x,y)$ provide a simple counterexample.

Permutation of Inferences.

Theorem 3 and its corollary are permutation theorems in the sense that they can be interpreted as stating that inferences in certain derivation trees can be permuted in some regular way. Theorem 5 and its corollary are permutation theorems in the same sense. The corollary to Theorem 5, stated in terms of paramodulation, asserts that Theorem 1 continues to hold with restrictions (r1) - (r5) where (r4) and (r5) are the following:

(r4) No resolvent is the parent of a paramodulant.

(r5) Given any complete resolution method for the predicate calculus (i.e. set-of-support, P_1-deduction, AM-clashes, etc.) we may insist that every resolvent be generated in accordance with that method.

The completeness of the method corresponding to Theorem 1 and restrictions (r1), (r2), (r4) and (r5) can be obtained by analyzing the abstract of the original Robinson-Wos completeness proof [16]. It was in fact this observation which originally motivated Theorem 5. Theorem 1 and restrictions (r3) and (r4) assert hat any paramodulation and resolution refutation can be obtained in the canonical orm where all trivialisations of inequalities precede all paramodulations which recede resolutions.

Suppose that $S = S_0 \dot{\cup} S_1$ and that each clause in S_1 is non-positive. Define he set of $\underline{S_1\text{-resolvents from } S_0}$ to be the smallest set containing

(1) each clause $C \in S_0$ and

(2) each resolvent of a clash with nucleus a complete factor of some clause in S_1 and with satellites factors of S_1-resolvents from S_0.

otice that each resolvent obtained by (2) is obtained by resolving on all and nly on the distinguished literals of the complete nucleus factor of the clash. n case $S_1 = E_2 \cup E_3$ each S_1-resolvent from S_0 is either a clause in S_0 or a clause btainable from $S_0 \cup E_4$ by primary paramodulation without resolution.

<u>Theorem 5.</u> If $S = S_0 \dot{\cup} S_1$ is unsatisfiable then some finite set S^* of S_1-resolvents from S_0 is also unsatisfiable.

<u>Corollary 2.</u> If the set of clauses S has no normal model and if S_0 is the losure of S under the operation of trivialising inequalities, then there is a inite unsatisfiable set of clauses S^* such that $C \in S^*$ implies that $C \in S_0$ or can be derived from $S_0 \cup E_4$ by paramodulation without resolution.

<u>Proof of Corollary 2.</u> If S has no normal model then $S \cup E$ is unsatisfiable nd, by Corollary 1, $S_0 \cup E_2 \cup E_3$ is unsatisfiable. Taking $E_2 \cup E_3 = S_1$, applying heorem 5 and the definition of S_1-resolvent from S_0, the conclusion of the coroll- ary follows.

The proof of Theorem 5 requires two lemmas.

<u>Lemma 1.</u> Let T be a hyper-resolution derivation of a positive non-tautologous ground clause C from ground clauses $S \dot{\cup} \{D\}$ where D is non-positive and occurs in T only at the nucleus node of T immediately above the root. Then there is a hyper-resolution derivation T^* of a clause $C' \subseteq C$ from clauses S^* where S^* is a set of $\{D\}$-resolvents from S.

<u>Proof of Lemma 1.</u> Let $D = \{\overline{L}_1,\ldots,\overline{L}_n\} \dot{\cup} D_0$ where D_0 is the maximal positive subclause of D; let $C = \{K_1,\ldots,K_m\}$. Then $S_0 = S \cup D \cup \{\{\overline{K}_1\},\ldots,\{\overline{K}_m\}\}$ is unsatisfiable since $S \cup D$ implies C. Let $S_0' = S \cup R \cup \{\{\overline{K}_1\},\ldots,\{\overline{K}_m\}\}$ where R is the set of resolvents of clashes $\mathcal{C}$ having clauses in S as satellites, D as nucleus and where the literals $\overline{L}_1,\ldots,\overline{L}_n$ are the literals in D resolved upon in $\mathcal{C}$. $S^* = S \cup R$ is a set of $\{D\}$-resolvents from S. We shall show that S_0' is un- satisfiable. By the unsatisfiability of S_0' it will follow that S^* implies C and by Theorem 4, since C is not a tautology, there is a hyper-resolution derivation T^* of a clause $C' \subseteq C$ from S^*.

To prove the lemma it remains to show that S_0' is unsatisfiable. Let T' be a binary semantic tree for S_0 relative to the set of atoms occurring in S_0

ordered in any way. Then T' is closed for S_0 since S_0 is unsatisfiable. It suffices to show that if $\mathcal{B}$ is a complete path from the root of T' to a tip of T' such that D fails at some node $N \in \mathcal{B}$ then some clause $D' \in R$ fails at some node $N' \in \mathcal{B}$. From this it follows that T' is closed for S_0' and that S_0' is unsatisfiable.

Let $\mathcal{B}$ be such a path and $N \in \mathcal{B}$ such a node. For $1 \leq i \leq n$ let $\mathcal{B}_i$ be the complete path of T' which differs from $\mathcal{B}$ only in L_i, i.e. $\mathcal{B}_i = (\mathcal{B} - \{L_i\}) \cup \{\overline{L}_i\}$. Then each such $\mathcal{B}_i$ contains a failure point N_i for some clause $A_i = \{L_i\} \dot{\cup} A_{0i} \in S_0$. But $A_i \neq D$ since $\mathcal{B}_i$ does not contain L_i and $A_i \neq \overline{K}_j$ for any j, $1 \leq j \leq m$, since L_i is positive and $\overline{K}_j$ is negative. Therefore $A_i \in S$. Let $D' = A_{01} \cup \ldots \cup A_{0n} \cup D_0$. Then D' is the resolvent of the clash $\mathcal{C} = \{A_1, \ldots, A_n, D\}$ and D' fails at some node $N' \in \mathcal{B}$.

<u>Lemma 2.</u> Suppose T is a hyper-resolution derivation of $\square$ from ground clauses S and suppose $C \in S$ is a positive clause which occurs at a tip $N \in T$. Let $C' \subseteq C$. Then there is a hyper-resolution derivation T' of $\square$ and a one-one correspondence Φ from the tips of T' onto a subset of the tips of T such that

(1) C' occurs at the tip $N_0' = \Phi^{-1}(N_0)$ in T', and

(2) for all tips $N' \in T'$, N' N_0', the clause occurring at N' in T' is identical to the clause occurring at $\Phi(N')$ in T.

<u>Proof of Lemma 2.</u> (by induction on the number n of nodes in T). If n = 1 then $C = C' = \square$, T' = T and the correspondence Φ is the identity. Suppose now that T has k nodes and that the lemma holds for any derivation tree having fewer than k nodes. Let $\mathcal{C}$ be the hyper-resolution clash of which C is a satellite at N_0. There are two cases to consider: (1) the L in C resolved upon in $\mathcal{C}$ occurs in C' and (2) L does not occur in C'.

<u>Case (1).</u> Let $\mathcal{C}' = (\mathcal{C} - \{C\}) \cup \{C'\}$. Then $\mathcal{C}'$ is a hyper-resolution clash and its hyper-resolvent D' subsumes the hyper-resolvent D of $\mathcal{C}$. Let T_1 be the derivation obtained by ignoring all of the nodes in T lying above the node N_1 which lies immediately below N_0. Then T_1 is a hyper-resolution derivation of $\square$ and the clause D at N_1 is subsumed by D'. T_1 contains fewer than k nodes and by induction hypothesis therefore there is a hyper-resolution derivation T_1' and one-one correspondence Φ_1 from the tips of T_1' onto tips of T_1 such that D' occurs at $\Phi_1^{-1}(N_1)$ and for $N' \in T_1'$, $N' \neq \Phi_1^{-1}(N_1)$, the clauses occurring at N' and $\Phi_1(N')$ are identical. Let T_0' be obtained from the subtree T_0 of T rooted in N_1 by replacing the clauses C and D at N_0 and N_1 by C' and D' respectively. Let $\Phi_0(N) = N$ for every tip N of T_0'. T_0' is a hyper-resolution derivation of D'. Let T' be obtained by identifying the tip $\Phi_1^{-1}(N_1)$ of T_1' with the root of T_0'. Then T' is the desired hyper-resolution derivation of $\square$ and the desired mapping Φ is defined as Φ_1 for tips of T' which belong to T_1' and as Φ_0 for tips of T' which belong to T_0'.

Case (2). If $L \not\in C'$ then C' subsumes the resolvent D of $\mathcal{C}$. Let N_1 and $\mathcal{C}_1$ be as in case (1). Since T_1 contains fewer than k nodes and since C' subsumes D, the induction hypothesis applies to T_1 and to the node $N_1 \in T_1$. The derivation T_1' and mapping Φ_1 such that C' occurs at $\Phi_1^{-1}(N_1)$ are the desired hyper-resolution derivation T' of $\square$ and mapping Φ.

Proof of Theorem 5. Suppose first that S is a set of ground clauses. Let S^* be the finite set of all S_1-resolvents from S_0. Let T be a hyper-resolution derivation of $\square$ from S containing no tautologies and let T' be obtained from T by consecutively deleting all nodes above each S_1-resolvent from S_0, i.e. T' is the hyper-resolution subtree of T which derives $\square$ from clauses in $S^* \cup S_1$ and which contains S_1-resolvents only at its tips. We shall transform T' into a tree T_0 which derives $\square$ from clauses in S^*. It will follow that the finite set S^* is unsatisfiable.

The construction of T_0 is by induction on the number n of occurrences of clauses in S_1 at the tips of T'. If $n = 0$ then T' is already a derivation of $\square$ from some subset of clauses in S^*. Suppose then that T' contains $k > 0$ occurrences of clauses from S_1 at its tips and suppose that any hyper-resolution derivation T'' of $\square$ from $S^* \cup S_1$ which contains fewer than k such occurrences and no tautologies can be transformed into a derivation T_0 of $\square$ from S^*. We shall transform T' into such a tree T''. Then T_0, the transform of T'', is also the desired transformation tree for T'.

Let N be an interior node in T' such that the hyper-resolvent C occurring at N is the resolvent of a clash with nucleus $D \in S_1$ and such that the tips of T' lying above N contain only this one occurrence of a clause from S_1. The subtree of T' rooted in N derives C from $S^* \dot{\cup} \{D\}$. By Lemma 1, since C is not a tautology, there is a hyper-resolution derivation T_1 of some $C' \subseteq C$ from S^*. Let T_2 be obtained from T' by ignoring all of T' above the node N. Then, by Lemma 2, there is a hyper-resolution derivation T_3 of $\square$ from $S^* \cup S_1 \cup \{C'\}$ and a one-one correspondence Φ from the tips of T_3 onto a subset of the tips of T_2. T_3 contains fewer than k occurrences of clauses from S_1 at its tips and the clause C' occurs at the tip $\Phi^{-1}(N)$ of T_3 corresponding to N in T_2. Let T'' be obtained from T_1 and T_3 by identifying the root of T_1 with the tip $\Phi^{-1}(N)$ of T_3. T'' is the desired hyper-resolution derivation of $\square$ from $S^* \cup S_1$. That T'' contains no tautologies can be verified by checking that the derivations T_1 and T_3 contain no tautologies.

If S is not a set of ground clauses then let $S' = S_0' \dot{\cup} S_1'$ be an unsatisfiable set of ground instances of clauses in S, where S_0' and S_1' are instances of clauses in S_0 and S_1 respectively. By the part of the theorem already proved, there is a finite unsatisfiable set $S^{*\prime}$ of S_1-resolvents from S_0'. By the lifting lemma for clashes, for every clause $A' \in S^{*\prime}$ there is an S_1-resolvent A from S_0

which has A' as an instance. Let S* be the set of all such A for all A' $\in$ S*'. Then S* is unsatisfiable since its set of instances S*' is unsatisfiable.

The reader familiar with Andrews' paper [1] will note the similarity between the proof of Theorem 5 using Lemmas 1 and 2 and the proof in [1] of Theorem 1 using Lemmas 1-5.

Concluding Remarks.

(1) The argument for using hyper-resolution with equality axioms is based on a comparison with paramodulation and resolution applied to sets of clauses containing the axioms E_4. In this connection it should be noted that Robinson and Wos [15] conjecture the completeness of a more restricted paramodulation system: in this system one adds to a set of clauses S_0 which has no normal model just the clause $\{x = x\}$ and applies paramodulation and resolution to derive $\Box$. Interpretation of this system in terms of hyper-resolution is not entirely straight-forward and comparison of these two systems is therefore correspondingly more difficult.

(2) The set E_2 need not include axioms for Skolem-function letters f which result in S_0 from the elimination of existential quantifiers. That this is so is easily verified by noting that before eliminating existential quantifiers we need only include axioms of functional substitutivity E_2 for the function letters actually occurring in the original fully quantified set of sentences. This improvement of the hyper-resolution method induces a corresponding improvement of (r1) and (r2) in the paramodulation method. In the case where the original quantified set of sentences contains no function letters, the set E_2 is empty, and for paramodulation, (r1) and (r2) state that E_4 may be replaced by the single clause $\{x = x\}$. We do not consider that the well-known procedure for eliminating function letters by introducing new predicate letters reduces the problem of proving the Robinson-Wos conjecture to the special case just verified. This conjecture remains an important problem which has counterparts in the f-matching method [4] , in the lifting lemma for generalised resolution [13] and in E-resolution [10] .

References.

[1] Andrews, P.B., "Resolution with Merging", Journal of the Association for Computing Machinery, 15, 367-381, July 1968.

[2] Brown, T.C. Jr., "Resolution with Covering strategies and Equality Theory", internal memorandum, California Institute of Technology, 1968.

[3] Chang, C.L., "Renamable Paramodulation for Automatic Theorem-Proving with Equality", internal memorandum, National Institutes of Health, Bethesda, Maryland, 1969.

4] Darlington, J.L., "Automatic Theorem-Proving with Equality Substitutions and Mathematical Induction", Machine Intelligence 3, edited D. Michie, Edinburgh University Press , 113-127, 1968.

5] Hayes, P.J., and Kowalski, R., "Semantic Trees in Automatic Theorem-Proving", Machine Intelligence 4, edited B. Meltzer and D. Michie, Edinburgh University Press, 1969.

6] Kowalski, R., "Studies in the Completeness and Efficiency of Theorem-Proving by Resolution", Ph.D. thesis, University of Edinburgh, 1969.

7] Lee, Char-tung, "A Completeness Theorem and a Computer Program for Finding Theorems Derivable from Given Axioms", Ph.D. thesis, University of California, Berkeley, 1967.

8] Meltzer, B., "Theorem-Proving for Computers: some results on resolution and renaming", Computer J. 8, 341-343, 1966.

9] Meltzer, B., "Some Notes on Resolution Strategies", Machine Intelligence 3, edited D. Michie, Edinburgh University Press, 71-76, 1968.

.0] Morris, J.B., "E-Resolution: Extension of Resolution to Include the Equality Relation", Proceedings of the International Joint Conference on Artificial Intelligence, Washington, D.C., 1969.

1] Robinson, J.A., "Automatic Deduction with Hyper-resolution", Int.J. Computer Math. 1, 227-234, 1965.

2] Robinson, J.A., "A Review of Automatic Theorem-Proving", Proceedings of Symposia in Applied Mathematics, 19, Mathematical Aspects of Computer Science, American Mathematical Society, 1967.

3] Robinson, J.A., "The Generalised Resolution Principle", Machine Intelligence 3, edited D. Michie, Edinburgh University Press, 77-93, 1968.

4] Robinson, J.A., "The Present State of Mechanical Theorem-Proving", Fourth Annual Systems Symposium, to appear in Proceedings.

5] Robinson, G., and Wos, L., "Paramodulation and Theorem-Proving in First-Order Theories with Equality", Machine Intelligence 4, edited B. Meltzer and D. Michie, Edinburgh University Press, 1969.

6] Robinson, G., and Wos, L., "Completeness of Paramodulation", Journal of Symbolic Logic, 34, 160, March, 1969.

7] Silbert, E.E., "A Machine-Oriented Logic Incorporating the Equality Relation", Machine Intelligence 4, edited B. Meltzer and D. Michie , Edinburgh University Press, 1969.

8] Slagle, J.R., "Automatic Theorem Proving With Renamable and Semantic Resolution", Journal of the Association for Computing Machinery, 14, 687-697, October 1967.

9] Slagle, J., Chang, C. and Lee, C., "Completeness Theorems for Semantic Resolution in Consequence Finding", Proceedings of the International Joint Conference on Artificial Intelligence, Washington, D.C., 1969.

0] Wos, L., Carson, D., and Robinson, G., "The Unit Preference Strategy in Theorem Proving", A.F.I.P.S. Conference Proceedings 26, Washington, D.C., 615-621: Spartan Books, 1964.

1] Wos, L., Robinson, G., Carson, D.F., and Shalla, L., "The Concept of Demodulation in Theorem-Proving", Journal of the Association for Computing Machinery, 14, 698-709, October 1967.

HILBERT'S PROGRAMME AND THE SEARCH FOR AUTOMATIC PROOF PROCEDURES

G. KREISEL

INTRODUCTION

Hilbert's programme concerned the formalization, that is mechanization of mathematical reasoning. It looked for formal languages (given by 'mechanical' rules as analysed by Turing) to represent mathematical assertions and for formal rules of inference to generate (representations of) mathematical proofs. As understood naively the search for automatic proof procedures, or more precisely, mechanical ones is the following *'practical' variant* of Hilbert's programme.

Not all meaningful mathematical assertions are to be considered but only *practical* or *feasible* ones; this is a limitation compared to Hilbert's programme. But also, and this is a sharpening, we are not now satisfied to settle an assertion by some formal derivation (according to given rules), but by one of practical or feasible complexity.

Since the pure mathematician is not accustomed to work with such concepts as *practical* or *feasible* which have not been formally analyzed for him, he tends to feel helpless and looks for ambiguities. But, as elsewhere in practical matters, there is no need and perhaps no hope of a complete analysis: one looks for properties (of the concepts above) which are both convincing and enough to settle the specific matter considered. I give a number of *examples* and *problems* in § 1 of such partial analyses which may be of intrinsic interest. Let us remember that Hilbert's programme, as he originally meant it, has failed, though it appeared plausible to somebody with his insight. In view of the obvious connection with automatic proof procedures, we must guard against similar misjudgements here since, with all due respect, people working in automatic proof theory cannot be expected to be superior to Hilbert. The best hope then, is *to look for, and make use of, something that we have and that Hilbert did not have!* for instance:

WORK ON HILBERT'S PROGRAMME

It so happens that quite recently (in [6]) I traced the exact sense in which Hilbert's programme failed, and the practical conclusions for proof theory that are to be drawn from this analysis. I now propose to look at the problem of automatic proof procedures in the light of the vast body of knowledge produced by work on Hilbert's programme. As Mr. Paul Getty's father (according to his successful son) always said: "No man's judgement is better than his information". Let us try to learn what we can, and avoid uninformed judgement.

What I have to say concerns a strategic conclusion about the *nature* of our problems, 'strategic' because it affects the general direction of research. The failure of Hilbert's programme shows that formalism does not provide a theory of the actual process of mathematical reasoning; *it is not,* so to speak, *a fundamental theory*. (Of course, formalization is useful, either in limited areas or in combination with other considerations.) But this switch alters the nature of the problems involved. If we have to do with (what we believe to be) a fundamental theory, it is silly to worry about applications; they will look after themselves. But if not, the *discovery of fruitful areas of application is more important and usually more difficult than the development of methods,* and the same applies to the discovery of the particular non-formal, that is non-mechanical, elements in combination with which formalization is useful. An example of the latter is what is called man-machine-interaction; here I'd like to contribute a little by being specific about what seems to be specially hopeful man-machine-interaction.

In § 2, I go into an analogue (for automatic proof methods) of the failure of Hilbert's programme in the strict sense; 'strict' because requirements are imposed on the selection or justification of the formal rules used. (Gödel's second incompleteness theorem is relevant to Hilbert's programme in the strict sense.) The conjecture is that a complete recursive proof procedure does not specialize to a feasible one when applied to theorems of feasible complexity.

In § 3, I go into the analogue of the failure of Hilbert's programme in its crude sense (established by Gödel's *first* incompleteness theorem). Here it seems easier to make a *positive* suggestion in connection with automatic proof procedures than in connection with Hilbert's original programme. Specifically, though we know the formal independence of certain well defined assertions (e.g. the continuum hypothesis which is second order determined) we do not have an effective proposal for deciding them. In the case of automatic proof theory we have an obvious candidate:

- physical, but non-mechanical methods.

('physical' in the sense of: according to the laws of existing physics.) I can say enough about this proposal to show that it is *non-trivial:* even in the cases where one shows that a *prima facie* non-mechanical procedure is in fact equivalent to a mechanical one, an idea is needed to establish this equivalence. Questions of this kind have intrinsic interest; the answers certainly tell us something about the character of existing physical theories.

At the end of the paper I raise a question which is basic when one discusses theorem-*proving* at all: how do we make precise the distinction between formal rules which are *computation* rules and those which are rules of *proof* ? Do computers help us to formulate the distinction? It is not touched, it seems, by the two principal points of the present talk, namely matters of length or feasibility and the possibility of physical, but non-mechanical procedures; more precisely these points distinguish between different computation rules and not only (or perhaps, not at all) between rules of computation and rules of proof.

1 - ORIENTATION : results and problems.

Obviously, as with any project, one can go on endlessly listing facts in support of the search for automatic proof procedures. To *test* an idea, we have to look for probable limitations. The obvious thing to do is to go over cases in mathematics where *surprising* methods were needed to solve a problem. In such cases there is at least a chance that there is no intelligible automatic solution. And, though intelligibility and length of proof are not *very* closely connected, we stand a chance of finding obstacles to automatic proof procedures *among* short theorems proved by surprising methods.

(a) To avoid confusion, let me begin with a *distinction.* Suppose we are given two formal proof procedures, say, in predicate logic: rules with and without cut. Then we have a crude measure, namely the lengths of the shortest derivations of a given theorem by means of the two methods; evidently, the shortest proof with an additional rule will not be longer than the shortest proof without it. This crude measure

is good for negative results, if one wants to show that a problem is *not* feasible. In positive results, a more realistic measure is needed: one considers the class of derivations *together* with an enumeration (the order in which the methods are tried out) and counts the trials and errors.

A natural (and, as far as I know, open) problem is to consider a genuine or 'deterministic' proof procedure involving proofs with cut, and to see how it compares for the second measure with the length of proofs without cuts. An interesting comparison in terms of the crude measure is given, in a slightly different context, by Dreben - Aandera - Andrews [4].

(b) As a first example of a 'surprising' argument we have: the irrationality of $\sqrt{2}$ or the proof of Fermat's conjecture for, say, exponent 3. I always had the impression that the surprising element was the use in the proof of notions which are not built up formally from the concepts used in stating the theorem proved. (For the most detailed discussion in print, see Shepherdson [10].)

I should have thought it likely that, for anything like a universal method, these very theorems would be long to prove, at least for the second measure in *(a)*. At any rate for a specific proof procedure my conjecture is a precise mathematical problem. Amusingly *this* mathematical problem can certainly be solved by an automatic method! One simply lets the method in question run on until one has reached the acceptable limit.

(c) The second example I whish to mention is more sophisticated. We know that delicate topological methods have been used by Adams [1] or Milnor [9] to show that

(*) there are no skewfields, or division algebras over the reals of dimension n for $n \neq 1,2,4,8$. Now, for any fixed n, say 256, the assertion (*) is a theorem in the language of the field of real numbers and therefore a consequence of the axioms for real closed fields.

The problem is this: are there n for which the formulation of (*) in the language of fields is feasible, but the formal derivation is not? For instance, if $n = 256$, and the shortest derivation is 2^{2^n} this would settle the matter.

Of course, the reason for such a situation would be that the decision method for real closed fields is simply not feasible (if one means here a method of assigning to each closed formula A a formal derivation of A or of $\neg A$, and not merely assigning to $\underline{A}$ its truth value $\top$ or $\bot$.

Naturally to find a suitable n one must think of one which 'barely' misses being a solution to (*)! There is a lot of talk about the need for abstract methods to make mathematics intelligible. Sometimes I get the impression that people are rather happy to think that this question will not be taken seriously because, allegedly, intelligibility is too hard to analyze. Well in *one* direction, we may substitute *length*, and if the length is really exorbitant we can be sure that there is no intelligible elementary proof. (This is quite unaffected by the existence of short unintelligible proofs.)

(d) A probably easier (but also less interesting) case to consider is in connection with Fermat's conjecture. I have heard it said (but this may have been myself) that for the *existing* computing machines *no counter example to Fermat's conjecture could be found by sheer computation*. To spell it out: the work of Vandiver [12] still leaves open the possibility that there are feasible x, y, z, n (that is printable using the usual representation of natural numbers) for which $x^n+y^n = z^n$; but, if so, one of the numbers x^n, y^n, z^n would be so large that *it* could not be printed.

Vandiver's results illustrate also another general point: they provide a *finite class* of natural numerical problems which we have solved by abstract methods but which, demonstrably, cannot be handled in a feasible number of steps by the *natural universal method*. To spell it out: it seems clear that, on minimal assumptions on feasibility, there are feasible quadruples (x, y, z, n) for which Vandiver establishes Fermat's conjecture, but x^n, y^n, z^n are not all feasible. In other words, for such (x, y, z, n), $x^n + y^n \neq z^n$ cannot be established by the natural universal method which consists in computing x^n, y^n, z^n from the recursion equations, then computing $x^n + y^n$, and finally comparing the result whith z^n.

I certainly don't want to suggest to anybody work which is only fit for coolies; but I have the impression that a clean *detailed proof* of this simple conjecture would be healthy for the whole subject of automatic proof procedures. After all, there is an immediate next step of establishing the conjecture for greater varieties of coding as in Winograd [13].

The examples above should be enough to indicate possible limitations of automatic proof procedures. I shall now go into the *positive* side, in particular consider the *combination* of automatic methods with other considerations.

(e) An example from number theory (that I happen to know). Without being logically spectacular, the following use of mechanical procedures, indeed of mechanical gadgets (a computer) seems noteworthy [8]. Using the usual notation of number theory

$$(\exists n < 1.65.10^{1165})\ (\pi n > li\, n)$$

is established. It is a 'theorem', not a pure computational result, because the result was not established by computing $\pi n - li\, n$ for all $n < (1.65)10^{1165}$, but as follows. First a relatively small bound for n is established on the assumption that there are no zeros $\sigma + it$ of Riemann's zeta function for $|t| < T_0$ and $\sigma > \frac{1}{2} + \sigma_0$. Next a bound $B(\sigma,t)$ for n is obtained if there is a zero for $0 < t < T_0$ and $\frac{1}{2} + \sigma_0 < \sigma < 1$, and *the bound happens to be bad if* $|t| < T_1$, $T_1 \ll T_0$. All this is spotted by a man. Here the machine takes over, and excludes zeros off the half line $\sigma = \frac{1}{2}$ for $0 < t < T_1$. (Perhaps a machine was even used to compute values of the formal expression $B(\sigma,t)$ which led to the choice of T_1.)

(f) A suggestion for non-numerical tasks for automatic computers. Let us here accept the claim that mathematicians have discovered relatively few important abstract(1) structures like topological spaces in terms of which mathematics becomes intelligible. Most spectacularly, problems concerning *finite* configurations, e.g. number of solutions of a polynomial in a finite field, become manageable when the configuration can be endowed with a suitable abstract structure. It seems reasonable to suppose that computers could be used with good effect in this area of research; specifically I expect we shall find situations in practice where there are just too many possible relations to try out by 'hand', but not too many for a computer to run through. If it is really true that it is marvellous to discover one of these abstract structures in a concrete finite configuration our computer could do a marvel beyond our power.

(1) *Correction.* In my lecture I exaggerated the difference between two uses of 'abstract': in logic where one means higher type notions, in modern mathematics where abstract notions such as groups are often defined by first order axioms (but not always; e.g. compact topological spaces). I overlooked the fact that in the example here used the first order notion of topological space was abstracted from higher type geometric notions.

2 - SELECTION OF MECHANICAL PROOF PROCEDURES

Both direct and circumstantial evidence suggests that people who are enthusiastic about the subject feel that the selection shouldn't be too difficult, just as Hilbert felt about the selection of formal systems. For his foundational interests, the sole criterion to be satisfied was that the consistency of the formal system considered should be provable by combinatorial methods, and he thought this shouldn't be too hard because the consistency of the systems actually considered was in fact evident, albeit on non-combinatorial grounds. What is the corresponding simple minded faith behind a (simple minded!) search for automatic proof procedures? I think it's this.

Suppose we already have a mechanical proof procedure, as in the case of predicate calculus (this corresponds to already knowing the consistency); the 'faith' is that in a natural way this will yield a ***feasible*** proof procedure for feasible theorems ('feasible' corresponds to Hilbert's 'combinatorial').

Conjecture:
Under reasonable conditions on feasibility[2], there is an analogue to Gödel's second incompleteness theorem, that is the article of faith above is unjustified.

Let's not get lost in trite speculations about the future, but let us organize what we know, namely the history of Hilbert's programme. The following points strike me as most significant. (There is hardly general agreement on this; but since most people have not worked successfully in proof theory their agreement would have little weight.)

(a) The positive and negative value of Gödel's second theorem. Not unnaturally, *within* proof theory, its (positive) value though real, is somewhat technical: it serves as a crosscheck on proposed consistency proofs, or as a means of comparing different formal systems. As everybody knows its negative impact was terrific; people dropped proof theory like a hot brick. One may well point out that, objectively, this was unjustified because what Gödel's theorem showed was that a more sophisticated formulation of Hilbert's programme was needed, not that it was

(2) Note that Gödel's original statement only applies to 'reasonable' formal systems in which, e.g., ***modus ponens*** is, more or less, a derived rule; it does not apply to certain cut free systems whose importance was not recognized at the time; cf. e.g. [6], footnotes 8 and 16 on p. 331 and p. 349 resp.

bankrupt. At the risk of sounding arrogant I find people's reaction perfectly sound. They had so evidently and totally misjudged the nature of their own project that they had no reason for believing in their own talent in this particular area of research; therefore it was perfectly reasonable for them to want to drop it: a burnt child keeps shy of the fire.

Clearly, people who were overimpressed by early minor successes of automatic proof procedures run a similar risk, and objectively exaggerated despondency may be quite sound from their own point of view.

Digression:

Just because people tend to get overexited by partial successes of an automatic proof procedure they do not seem to analyze why it works; as if they were frightened of the result of giving the matter a second thought. This reminds me of a similar attitude in connection with finding effective bounds which is not directly connected with the present topic, but may be of interest to some readers.

Thue - Siegel - Roth - Baker theorem

Let α be an algebraic number of degree n (given by its defining polynomial), let all *l.c.* variables range over natural numbers. For various functions f it was proved that

$$\exists m\, \forall q\, \forall p\, (q \geq m \rightarrow |\alpha - p/q| > q^{-fn}), \text{ which we write}$$

$$\exists m\, \forall p\, \forall q\, A(m,p,q)\ .$$

Thue, Siegel, Roth for: $fn = \frac{1}{2}n$, $\sqrt{n}$, $2 + \varepsilon$ (any $\varepsilon > 0$) *without* determining m (as a function of α, resp. α and ε), Baker for a much larger fn, but with explicit determination of m. Incidentally, we do not know if there is a *recursive* function m of α even in the case $fn = \frac{1}{2}n$.

Davenport and Roth [3] showed that even for $fn = 2 + \varepsilon$ one *can* explicitly determine a bound, not for the *size* m on the exception, but for the *number* of q for which $|\alpha - p/q| \leq q^{-fn}$. Is this a mere curiosity?

No. As so often, in Roth's proof the non-constructive part merely involves predicate logic, specifically the law of the excluded middle.

By Herbrand's theorem we have functions μ such that

$$A(\mu_0,p,q) \vee A[\mu_1(p,q),p_1,q_1] \vee \ldots \vee A[\mu_\kappa(p,\ldots,p_{\kappa-1}, q \ldots,q_{\kappa-1}),p_\kappa,q_\kappa]$$

is a consequence of purely numerical identities (from which the non-constructive argument starts). Inspection shows that all the μ are uniformly bounded by μ_0 say, and then it is clear that there cannot be more than k exceptions $\bar{p}_i/\bar{q}_i$ $(0 \leq i < k)$ exceeding μ_0. If we replace (p,q) by $(\bar{p}_0,\bar{q}_0)$ and the *variables* (p_i,q_i) by the numbers $(\bar{p}_i,\bar{q}_i)$ all disjuncts are false except, possibly, the last one. Consequently

$$\forall q_k \; \forall p_k \; (q_k \geq \mu_0 \rightarrow |\alpha - p_k/q_k| > q_k^{-fn}) \, .$$

Note that we do not even have to determine μ_0 explicitly.

Thus Herbrand's theorem for $\exists\forall$ formulae together with the empirical fact that non-constructive proofs in number theory rarely apply induction to logically complicated formulae gives an explanation of the success of [3]. Here ends the digression.

(b) In what *terms* can we hope to select automatic proof procedures? Or rather, since I don't want to slip into speculation, *in what terms were successful selections of formal rules made?*
I believe one cannot exaggerate the importance of the fact that the selection was extremely sophisticated. Even speaking purely formally, that is from the point of view of syntactic properties, Heyting's systems are outstanding; they were found by the requirement that the rules be valid for the (highly abstract) notion of intuitionistic proof. Another formally outstanding class of formal rules are Gentzen's so-called cut free rules which were dictated by the requirement that a formal derivation of a theorem A should exhibit possible truth conditions on the parts (subformulae) of A. Thus the actual selection principle did not use formal criteria such as duality (which turn out as by-products) but abstract interpretations of formal derivations. For the corresponding but simpler situation in the selection of axioms, see [6] *passim*, p. 332 *(b)*, p. 360 etc. I believe the question of the selection of formal rules is much neglected. Without settling the matter of cause and effect, it should be noted that *formalist foundations suppress this question as a matter of principle.* They emphasize that a formal derivation (from mechanically listed axioms by means of

mechanical inference rules) has the same (mechanical) character as a *computation*. While the observation is perfectly true, the emphasis would be simply silly if one recognized at the same time that the selection of rules of *proof* is derived by means of *abstract* considerations. (In addition one has to distinguish them from the abstract considerations used in obtaining *computation* rules; cf. end of introduction and of § 5.)

(c) What could be done in proof theory after Gödel's second theorem and *before* current work on the selection of formal rules in terms of abstract notions such as intuitive concepts of proof? (This is the present situation with automatic proof procedures in terms of the analogy pursued in the present paper; the analogue to current work in proof theory would then be an analysis of the distinction between computation rules and rules of proof mentioned above and taken up in §5.)

Looking back at my own reaction, I think it was pretty sensible (for me at least). Feeling uncertain about the logical significance of the rules considered, I looked for their 'mathematical significance' that is, for *ad hoc* consequences (formulated in terms of ordinary mathematics) which were of simply obvious interest. In this way one learnt a lot of facts about the rules themselves too. The examples and problems in § 1 are intended to be of the same general kind.

Discussion of two procedures of research

The first is *ad hoc*, as described above. Naturally it goes *round* the basic issues instead of *facing* them, since one is trying to find out by means of typical examples which issues are basic. An alternative is to experiment systematically with different formulations of the issues which are illustrated by the examples. What are the relative merits of the two methods? When beginning a new field of research I favour the *ad hoc* method: an example is sufficient for a negative result and also people tend to get absorbed in systematic work for its own sake and lose sight of the practical problem. An area where, I believe, my own preference is supported by experience is related to *(b)* above, namely the selection of restricted rules of inference by topological or lattice theoretic interpretations of the 'logical' operations (a selection principle in *familiar* abstract, in fact, algebraic terms). Of course it's hard to distinguish in an analysis of progress between the role of a method and the role of the people using it. But the fact is that the algebraic approach always limped behind, doing predicate logic when striking results on arithmetic were available etc. I think this should be borne in mind, even though the algebraic approach may pay off in future.

3 - ANALOGUE VERSUS TURING COMPUTERS

Let me consider consequences of the following conjecture corresponding to Gödel's *first* incompleteness theorem.

Under reasonable conditions on a class of theorems (including, for instance, that the class is not recursive) there is no mechanical procedure at all for deciding each feasible theorem in a feasible number of steps.

The strongest result one could expect is simply that the *complexity* of the sequence $(\pm)_p$ for a *single* formula $A(p)$ with parameter p for evidently feasible p is evidently not feasible where $(\pm)_p = T$ if $A(p)$ is true, $(\pm)_p = \perp$ if $A(p)$ is false. For an example, for Ehrenfeucht's notion of complexity in terms of k-computability, see the dissertation [11]. (Rabin is, however, dubious about the measure.)

Now, one reaction to a proof of this conjecture could be that of ordinary mathematicians to Gödel's first theorem, namely, practically speaking, to ignore it; they leave (mechanically) unsolvable problems unsolved, and occupy themselves with Riemann's hypothesis, Poincaré's conjecture and other worthwhile problems. Correspondingly one might look for efficient automatic proof procedures in *suitably selected areas*.

Digression : The high standard of work in mathematics should be compared with corresponding work in logic, by contrasting 'solvable cases' in predicate calculus and in diophantine equations. The correspondence is reasonable since not knowing whether Hilbert's tenth problem is solvable is not too different, practically, from knowing that it is unsolvable.

The first step is a crude classification, e.g. one considers equations in *one* variable and arbitrary degree, or equations of *degree* ≤ 2 and an arbitrary number of variables. But after that, instead of continuing with this simple minded measure in terms of *prefix* (that is number of variables) or *matrix* (according to degree only) one uses classifications that cut *across* this measure. The theoretical considerations that led to the more fruitful classifications are sometimes delicate and some-

times not even explicit.
The proper analogue in predicate logic is a classification by *mathematical content* (instead of prefix and formal properties of the matrix). After all, the best 'solvable cases', though they do not usually go by this name, concern classes of formulae of the form

$$A \longrightarrow F$$

where A is a *decidable theory* and F an *arbitrary* formula in the language of A.

I think this simple point is worth making as illustrating what kind of selection principle may be fruitful. (Here ends the digression.)

Another reaction is to see *whether the physical universe could be of help.* Note that, e.g. in § 2, there was no mention of physical devices, we talked of the mathematical properties of *rules*, not of their physical realization. Now we look at computers engaged in machine proving as follows.

We believe we know enough about the physical properties of the computer to assume that, with high probability, it does what it is designed to do, namely to realize certain formal rules. The fact that there is a mechanical check is not *used* because the whole point of a fast computer is that we shall not even try to apply this check; after all, the whole point of a computer is that (at least occasionally) it will be useful to *know* certain results without knowing their proofs.

With the *same* assumption, we may appeal to the whole body of physical knowledge, and ask what sequences can be generated by physical devices *according to accepted theory* with roughly the same probability. Or rather, to bring in feasibility, at what *cost* in production and time of operation, sequences can be generated.

The *characterization* of the sequences is given in the language of current physics, which certainly includes the language of arithmetic. *It is therefore not yet excluded that some of the sequences in question are non-recursive.* (Monumental work on the theory of partial differential equations was needed to show that the *bulk* of systems following the laws of the classical theory of continuous media, do have recursively approximable behaviour. But there are plenty of open problems in the quantum theory.)

The idea is perfectly parallel to using physical considerations as a help for studying mathematical problems in the theory of partial differential equations. For example, Dirichlet's principle is got from the assumptions that *(i)* certain parts of theoretical physics and *(ii)* certain qualitative impressions we have of the universe are exact. Obviously this is not a mathematical proof; but there is certainly no doubt of the heuristic value of these assumptions. What I have in mind here is a *theoretical investigation* whether, according to present physical theory, all physically realizable sequences are recursive and, more sharply: whether analogue computers are reducible to Turing machines preserving feasibility.

The reader will find a discussion of this whole matter in [5]. Just because it was published some years ago, it seems proper to report what has happened since then.

(a) According to the best information I have been able to obtain my specific suggestion l.c., p. 270, concerning *size* of molecules has not been refuted, but it is not plausible. Precisely: are there large molecules whose spectrum (or: to have a dimensionless quantity, the ratio of the first spectral line to the second) is not recursive? The general impression is that *Kato's* theorem could be used to give arbitrarily close recursive approximations; similarly for spins.

(b) Another specific suggestion, properly formulated in [7] footnote 2, concerns the three body problem: for any neighbourhood U in phase space, time t and any n can we determine *recursively* neighbourhood V such that *(i)* $|V - U| < 1/n$ and there is a collision before $t + n^{-1}$ for *some* position in V or *(ii)* $U \supset V$ and there is no collision before $t - n^{-1}$ for every position in V. Though the set of neighbourhoods which do not lead to collisions before t is *recursively enumerable* the latest information that I have been able to obtain leaves open if it is *recursive*[3].

(3) The paper [7] quoted above develops thoroughly the discussion in [5], pp. 269-270 concerning intuitionistic mathematics, including Kripke's refutation of Church's thesis from the axioms I gave for the 'thinking subject'.

(c) The argument in footnote 1 on p. 267 of [5] can be extended to stochastic processes with an *infinite* number of discrete states and a recursive table of transition probabilities if the following definition (to be discussed in a moment) of a *sequence of states with non-zero probability* is used.

Let μ be the probability which the given table assigns (unambiguously) to *finite* sequences of states, and f_n the initial segment of length n of the infinite sequence f. For $n \leq m$, if c_m is a sequence of length m, c_n denotes its initial segment of length n. Then, we say:

$$(f \text{ has probability } > p^{-1}) \Longleftrightarrow \forall n \, \exists m \, [\mu\{c_m : f_n \equiv c_n\} > p^{-1}].$$

Such an f must be *isolated*, in fact there cannot be p distinct functions with this property altogether. Furthermore m can be found *recursively* from n because we need only finitely many c_m such that $\mu\{c_m : f_n \equiv c_n\} > p^{-1}$, a probability measure being monotone.

This argument supersedes footnote 1 on p. 267 of [5] where I used the 'accidental' additional fact that in the case of *finite* states, if f is isolated it is recursive; for infinitely many states one can conclude *immediately* only that f is hyperarithmetic.

The definition above was used, not to ensure the answer (which I didn't want at all) but to be sure that the values of f can be effectively determined from observations. I suspect I defeated my own purpose by requiring too much effectiveness in the manipulation of the observations.

4 - <u>REMARKS ON THE MALAISE</u> created by (claims made for) the search for automatic proof procedures. I think Rabin expressed it very well in his concluding remarks.

Mathematics is a particularly difficult part of intellectual behaviour; so one should not expect it to be the first of such areas to be made automatic.

Let us see how the exposition above fits in with his point.

(i) As long as we are looking for *limited* areas of application, as in § 1 and elsewhere in this talk, the point does not seem to have much weight. After all, we know already that fast computers (leave aside

theorem-proving machines) *replace* and *surpass* our intellectual powers in a quite realistic sense: even if we allow ourselves to use *all* mathematical methods at our disposal, computers generally multiply large numbers faster and more accurately than we. This same example shows that what is difficult for us is not necessarily particularly difficult to mechanize, and conversely. More generally, from a severely practical point of view the *differences* between human and mechanical mathematics are most important; it is they, by and large, which give us hope that we shall surpass human performances by mechanical means.

(ii) If, however, one proposed to make the *whole* of mathematics automatic (which amounts to treating automatic proof procedures at least as a candidate for a possible *mechanism* of mathematical reasoning) his point seems to me (still, cf. [5], p. 271, l. 1-5) very convincing.

Here, and in contrast to *(i)*, by accepting Rabin's point we lose the *glamour* of providing a theory of actual reasoning. (It's not hot news to claim: I'll build you a machine that is *different* from mind.) But I don't think we lose anything of value; on the contrary, we might even calm the 'backlash' of sensible, if less vocal people who are disgusted by vulgar exaggerations.

I do not mean to suggest that, at the present time, there is no hope of a worthwhile theory of mathematical reasoning; it's just that the advent of computers has not introduced anything essentially new over formalization itself. This leads me to a basic open question.

5 - RULES OF COMPUTATION VERSUS RULES OF PROOF

I am struck by a certain air of unreality in discussions of automatic proofs because the distinction between computations and proofs is not considered enough. Nobody denies the possibility of automatic computation anyway; but socalled proofs found by computers (of some algebraic identities) are at best on the border between computations and proofs. I think the distinction is best considered independently of computers altogether.

The single most important property of computers is that they realize certain formal instructions;

no recondite physics is involved. So inasmuch as they can be used in the theoretical analysis of anything, the formal instructions which they realize, do the job just as well. If we are astonished that computers can do a certain intellectual task, we should already be astonished (as well we might) that we have a formal analysis of this task. At best the computer is of practical use if the formal analysis is tedious and the computer is needed to check its consequences. For instance we may propose a strategy for playing chess but without a computer we'd never know what moves follow from this strategy; in other words we may need a computer to compare a formal theory with experiment. (It seems to me that this point was involved in several discussions, but not taken up.)

In contrast, ever since Turing, the literature introduces the picture of a computer into discussions of reasoning. A standard opening is an invitation to imagine a man in one room, and a computer in another (instead of simply requiring the 'other' man to follow mechanical instructions). The conclusion generally is that we shall not be able to distinguish the man from the computer by his responses (to a particular class of questions).

This is, of course, supposed to prove the 'similarity' of man and computer. But what is new about this? The old Greeks pointed out the similarity between the visible results of a man who understands what he is saying and one who is merely repeating the words. It would not be worth discussing misleading journalism, if it did not hide a genuine question.

Let us forget man, machine or any other physical system which realizes given formal rules and let us compare two formal derivations (assuming the intended interpretations to be given):

(i) an application of a deterministic system of recursion equations to *compute* a numerical function,

(ii) a derivation of induction up to $\omega^{\omega^{\omega^{\omega}}}$ in first order arithmetic (with a free predicate variable) which quite obviously was found by thinking out a *proof*.

There is a significant difference between these two objects: can we formulate what is essential about it? Is it that *(ii)* could, in a practical sense, only be found by means of a proof? Evidently useful as machines are for computation or data-processing one will do well to think up these questions before going on too far with 'theorem-proving'.

Remarks

I realize quite well that an abstract property or distinction may often be easier to analyze when one looks at *particular* realizations (for instance, one may find it easier to think clearly about the idea of simultaneity when considering the behaviour of fast moving objects). But as far as I can see, at least for an understanding in the natural practical sense of the word, in the present case a computer does not present anything new in principle over the formal rules which it realizes.

I realize, of course, also that though computers are theoretically irrelevant here, for psychological or sociological reasons they may provide the occasion for some logician to give a *perfect refined analysis far beyond the demands of practical understanding;* such an analysis, for instance 'computerlogic', would then help in the organization of information in cases which otherwise are wholly inaccessible: an instance of a change in degree (of refinement) leading to a difference in kind. I do not know of such work. Perhaps, the situation should be compared to formulations of basic issues, discussed at the end of §2.

B I B L I O G R A P H Y

[1] J.F. Adams, Vector fields on spheres, Ann. of Math 75(1962) pp. 603-632.

[2] A. Baker, Contributions to the theory of Diophantine equations, Phil. Trans. Roy. Soc. A 263 (1968) pp. 173-208 and, for practical bounds,

Linear forms in the logarithms of algebraic numbers (IV), Mathematika 15 (1968) pp. 204-216.

[3] H. Davenport and K.F. Roth, Rational approximations to algebraic numbers, Mathematika 2 (1955) pp.160-167.

[4] B. Dreben, P. Andrews, S. Aandera, False lemmas in Herbrand, Bull. Ann. Math. Soc. 69 (1963) pp. 699-706.

[5] G. Kreisel, Mathematical logic : what has it done for the philosophy of mathematics ? Bertrand Russel : Philosopher of the century, Allen and Unwin, London 1967, pp. 201-272.

[6] Survey of proof theory, J.S.L. 33 (1968) pp. 321-388.

[7] Church's thesis : a kind of reducibility axiom for constructive mathematics, to appear in the Proceeding of the conference on proof theory and intuitionism, Buffalo 1968.

[8] R.S. Lehman, On the difference $\Pi(x) - li(x)$, Acta arithmetica 11 (1966) pp. 397-410

[9] J. Milnor, Somme consequences of a theorem of Bott, Ann. of Math. 68 (1958) pp. 444-449.

[10] J.C. Shepherdson, Non-standard models for fragments of number theory, The theory of models, North Holland, Amsterdam, 1965, pp. 342-358.

[11] D.B. Thompson, Dissertation, Stanford University, 1968.

[12] H.S. Vandiver, Fermat's last theorem. Its history and the known results concerning it, Am. Math. Monthly 53 (1946) pp. 555-578 and

A supplementary note to a 1946 article on Fermat's last theorem, ibid.60 (1953) pp. 164-167.

[13] S. Winograd, How fast can computers add ? Scientific American 1968- (n°) p. 93-100.

A LINEAR FORMAT FOR RESOLUTION

D. W. Loveland

ABSTRACT

The Resolution procedure of J. A. Robinson is shown to remain a complete proof procedure when the refutations permitted are restricted so that clauses C and D and resolvent R of clauses C and D meet the following conditions: (1) C is the resolvent immediately preceding R in the refutation if any resolvent precedes R, (2) either D is a member of the given set S of clauses or D precedes C in the refutation and R subsumes an instance of C or R is the empty clause, and (3) R is not a tautology.

This work was supported by the Advanced Research Projects Agency of the Office of the Secretary of Defense (F44620-67-C-0058) and is monitored by the Air Force Office of Scientific Research. This document has been approved for public release and sale; its distribution is unlimited.

This research was also partially supported by NSF Grant GP-7064.

A Linear Format for Resolution*

Following the introduction of the Resolution principle as a complete strategy for demonstration of the inconsistency of an unsatisfiable set of first order clauses in Robinson [1965a], there have been several papers demonstrating restrictions on the generation of resolvent clauses while maintaining the completeness condition. Papers of this type include Robinson [1965b], Wos, Robinson, Carson [1965], and Andrews [1968]. In this paper also a restricted format for resolution is shown to be a complete strategy.

We assume familiarity with the notation and results of Robinson [1965a], in particular sections 2 and 5. Our concern is to deduce a contradiction from a finite set S of clauses. Each clause is itself a set of literals. Resolution may be taken as an operation mapping two parent clauses B and C into a resolvent clause D. If B and C are ground clauses and $L_1 \in B$ and $L_2 \in C$ are complementary literals then the ground resolvent of B and C is the set $(B - \{L_1\}) \cup (C - \{L_2\})$. The resolvent of arbitrary clauses B and C requires in general suitable instantiations of clauses B and C followed by the operation shown for ground resolution. The literals of B and C which under instantiation form the complementary literals are recorded in the key triple defined by Robinson.

A particular distinguished clause is the empty clause, denoted by $\square$. A deduction of clause C (from the set S) is a finite sequence $B_1, B_2, \ldots, B_n$ of clauses such that (i)B_i, $1 \le i \le n$ is either in S or a resolvent of B_j and B_k, $1 \le j, k < i$ and (ii) B_n is C. A refutation of the set S of clauses is a deduction of $\square$ from S. We define a linear deduction of C from the set S of clauses as a deduction of C from S such that $B_1, \ldots, B_k$ are in S and every B_i, $k+1 \le i \le n$ is a resolvent with B_{i-1} as one parent clause of the resolution. Each B_i, $i=k,\ldots,n-1$, is called a near parent clause. The other parent clause for B_{i+1} may be any B_j, $j \le i$. The subsequence $B_1, \ldots, B_k$, which serves to introduce the needed members of S

*This research was partially supported by NSF Grant GP-7064, ARPA #F44620-67-C-0058.

into the deduction, is called the prefix of the linear deduction. A linear refutation of the set S of clauses is a linear deduction of □ from S.

In section 7 of Robinson [1965a] the notion of subsumption is introduced. We state the definition here so as to include subsumption by the empty clause: given two distinct clauses B and C, B subsumes C precisely if an instance of B is a subset of C, i.e. $B\sigma \subseteq C$ for some substitution σ.

An s-linear deduction of clause D from the set S of clauses is a finite sequence $B_1, B_2, \ldots, B_n$ of clauses such that

(i) the sequence is a linear deduction of D;

(ii) if $B_1, \ldots, B_k$ is the prefix of the deduction and if $k+1 \leq i \leq n$ then one parent clause of B_i is either (a) from S or (b) a clause B_j, $j < i-1$, chosen so that the resolvent B_i subsumes an instance of B_{i-1}.

(iii) no tautology occurs in the sequence of clauses.

(A clause is a tautology if it contains complementary literals.) We shall prove the following theorem.

Theorem. The set S of clauses is unsatisfiable if and only if there is an s-linear refutation of S (i.e. an s-linear deduction of □ from S).

In an s-linear deduction of D, if $i > k$ then we shall call the parent clause of B_i which is constrained by condition (ii) an imported (parent) clause. We may slightly weaken condition (ii) to make more explicit the nature of the constraint on the imported clause. We note that for k as above, if $i > k$, then the imported clause C for B_i is either a member of S or has the property that there exists instances $C\sigma$ and $B_{i-1}\gamma$ such that for each literal L of C not appearing in the key triple $L\sigma \in B_{i-1}\gamma$. For a deduction consisting of ground clauses (a ground deduction) condition (ii) requires that the imported clause C is either in S or if L_1 is the literal in C that "disappears" in the resolution of B_{i-1} and C, then $C - \{L_1\} \subseteq B_{i-1}$.

The reader should note that it is not always possible to deduce □ from a given unsatisfiable set S of clauses if resolution is restricted by the requirement that one parent clause always be from S. If S is formed from the full conjunctive form on two predicates (i.e. $S=\{\{P,Q\}, \{-P,Q\},\{P,-Q\},\{-P,-Q\}\}$) we observe the only new clauses generated under the above constraint are four one-literal clauses plus two tautologies.

(If S is formed from the full conjunctive form on three predicates, then not even complementary one-literal clauses are derivable from S under the above constraint.) Restriction of one parent clause to membership in S hence does not produce a complete refutation strategy for resolution. Condition (ii) is a slight weakening of the "one parent from S" restriction, a weakening that is sufficient to allow completeness.

What is the purpose of studying such restrictions on the resolution operation? One reason, of course, is to obtain a better understanding of the concept of resolution. More practically, it is hoped that restrictions will trim the number of resolutions performed in the search for a refutation when attempted by hand or by computer. Unfortunately, it seems that with at least some of the restrictions already tested that the shortest refutation is often eliminated by the given restriction. Then the search for the longer refutations usually proves nearly or totally as big as the original search in spite of the reduced number of resolutions needed to consider all required deductions of a fixed length. Establishing the completeness of a restricted form of resolution is useful, however, in that any relaxation of the restriction need be considered only if it justifies itself by frequently realizing sufficiently shorter refutations. For example, it might develop in practice that neglecting the linearity condition is better than using it. That is, perhaps in practice one obtains a good strategy by insisting that every resolution have one parent clause taken from S or else that one parent clause "subsumes" the other parent clause as stated in the weaker version of condition (ii). Although such a strategy is complete because all s-linear deductions may be developed, it might happen that few of the refutations which appear first in a computer realization of the strategy happen to be linear.

Another strategy which is shown to be complete by the theorem is one closely related to that given in Andrews [1968]. Following Andrews, we say a <u>merge</u> of clauses B and C exists if there exists an instantiation $B\gamma$ of B and $C\delta$ of C such that a resolvent exists and $B\gamma \cap C\delta$ is non-empty. From the theorem stated earlier, it follows that S is unsatisfiable if and only if there exists a refutation including only resolvents with one parent clause either in S or a one-literal clause or the resolvent itself is a resolvent with a merge. It should be noted that this strategy differs somewhat from that of Andrews [1968] largely in that Andrews uses a merged resolvent as one criterion for a parent of an acceptable resolvent.

Hand calculation of a few simple examples leads one to surmise that when s-linear deductions are employed "depth-first" rather than "breadth-first" searches may be desirable. The s-linear deductions obtained on the attempted examples were in general longer than the unrestricted deductions, but were also easily discovered. This suggests the possibility that good planning heuristics can estimate the clauses in S likely to be needed so that few attempts (of quite some depth) are needed before an s-linear refutation is found. Question-answer systems seem one area where this approach may be desirable.

We turn our attention to the proof of the theorem. We make use of the basic Lemma of Robinson [1965a]. We paraphrase the summary statement of Robinson [1965b]. <u>If clauses</u> B <u>and</u> C <u>have instances</u> B' <u>and</u> C' <u>with resolvent</u> D' <u>then there exists a resolvent</u> D <u>of</u> B <u>and</u> C <u>with instance</u> D'. By induction it follows that if S is a set of clauses, if S' is a set of ground clauses, each clause of which is an instance of S, and if there exists a deduction of ground clause D' from S', then there exists a deduction of a clause D from S where D' is an instance of D. If D' is the empty clause, then D must also be the empty clause. Thus to show the existence of a refutation of S, it suffices to show the existence of a ground refutation from a suitable S'. Moreover, in section 2 of Robinson [1965a] it is shown that precisely if S is unsatisfiable, there exists a finite set S' of instances of S for which a ground refutation of S' exists. (Also see summary in Robinson [1965b]). These results allow us to establish the theorem at the ground level and obtain the full theorem by appeal to the stated results. (Care must be taken that the necessary distinctions in the definition of an s-linear deduction in the ground and general cases are correctly drawn. This will be left to the reader to verify; the translation is quite direct.)

It is immediate that if there is an s-linear refutation of S then S is unsatisfiable, due to the soundness of resolution. We must establish the converse. From the preceding paragraph it is clear we may assume from the unsatisfiability of S that a ground refutation of S' exists where S' is a finite set of instances of clauses of S. We need show the existence of a ground s-linear refutation of S'. For convenience we identify S' with S hereafter and consider all clauses of S to be ground clauses. We shall let $A, A_i, i=1,2,\ldots,$ denote atoms and $L, L_i, i=1,2,3,\ldots,$ denote literals. Certain early alphabet capital letters, perhaps with subscripts, shall denote clauses; occasionally $S_i, i=1,\ldots,m$ shall denote the m (ground) clauses comprising S. A ground resolution is conveniently pictured by use of a directed graph consisting of a one node tree. For example, if B and C are clauses and $L_1 \in B$ and $L_2 \in C$ are complementary literals with common atom A, a graph representing the resolution of B and C is given in Figure 1. We associate each parent clause with an incoming directed line segment and associate the resolvent clause D with the outgoing directed line segment. We associate the atom A with the node itself and label A the canceled atom of the node and of the resolution. The clause D, i.e. the set $(B - \{L_1\}) \cup (C - \{L_2\})$, does not have a literal with atom A if neither B nor C is a tautology.

Using the one node graph as a building block, we can display a refutation of S by a tree structure. Those clauses which are both resolvent clauses and parent clauses will label directed line segments passing from the node of the resolvent which formed the clause to the node of which it is one parent. The one outgoing line segment not pointing to a node, the final segment, is labeled by the empty clause; each incoming directed line segment not coming from a node, an initial segment, is labeled with a clause from S. Our assumption asserts the existence of such a tree. Figure 2 illustrates the tree giving the refutation of the set $S = \{\{P,Q\}, \{-P,Q\}, \{-Q\}\}$. Similarly, we can associate a tree structure with a deduction of clause D from S. Such a tree is called a deduction tree of D from S (or a refutation tree of S if D is the empty set). We shall often use the phrase "deduction tree of D" when S is determined by context. A minimal deduction tree of D is a deduction tree of D for which no collection of directed line segments and nodes can be removed so that (perhaps with relabeling) a new deduction tree of D from S is formed.

The directed line segments and nodes (and their labels) on a path from an initial segment to the final segment is called a branch of the tree. A branch is considered an ordered collection of directed line segments, nodes, clauses and canceled atoms with the order coinciding with the direction of the directed line segments, e.g., clause D (and the associated final segment) is last in the ordering. Phrases such as "node N_1 precedes node N_2 on the branch" refer to this ordering. There will be occasions when a

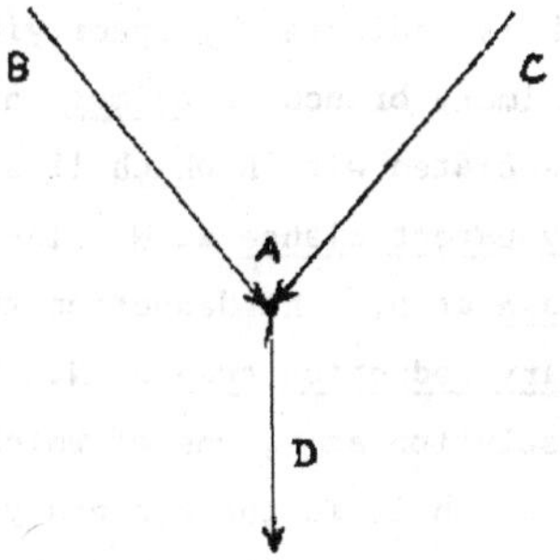

Figure 1

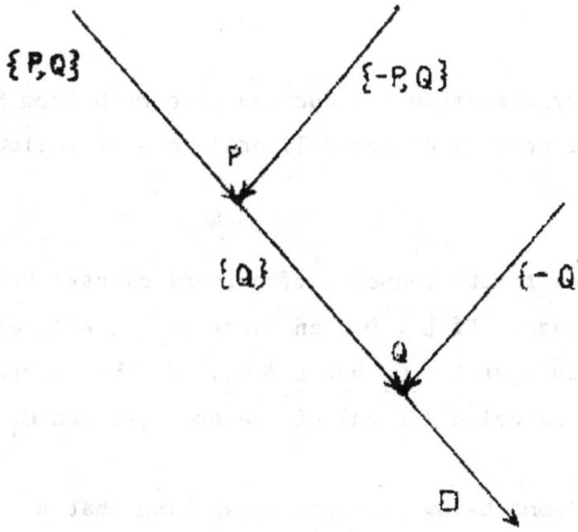

Figure 2

distinguished primary branch is indicated by specifying the initial segment. At a node N on the primary branch, a primary node, the parent clause of the resolution associated with N which lies on the primary branch is called the primary parent clause at N. The other parent clause is the secondary parent clause at N. The deduction tree of the secondary parent clause is the secondary deduction tree at N. The two complementary literals which allow the resolution at N, one of which is in the primary parent clause and the other which is in the secondary parent clause, are called canceled literals at N, one literal called the primary canceled literal and the other the secondary canceled literal. Clearly, both literals contain the canceled atom at N.

It is often useful to view a given deduction tree of D from S with a given primary branch as a sequence of primary clauses, the succeeding primary clause arising from a given primary clause by removal of one literal (the primary canceled literal) and the possible addition of new literals (from the secondary parent clause). In this regard we note the following fact.

Fact. If C is a given primary clause in a deduction tree of D from S and $L \in C$, then either L is a primary canceled literal in some following primary clause or $L \in D$.

The fact follows as there is a finite sequence of primary clauses between C and D, the last primary clause. If $L \notin D$ then there exists adjacent primary clauses C_1 and C_2 such that $L \in C_1$ but $L \notin C_2$. By the remark above, L must be the primary canceled literal of the node between C_1 and C_2.

The proof of the first lemma below proceeds by showing that a certain deduction tree is not minimal. We pass to a smaller tree structure by the operation of "removing a node N". The phrase remove (primary) node N shall imply the removal of all parts of the tree associated with primary node N, i.e., the secondary deduction tree at node N, the node N itself and the directed line segment of the resolvent at node N (with all associated labels). The primary parent clause G of node N becomes the primary parent clause of the following primary node N'.

The succeeding directed line segments on the primary branch must be relabeled with the correct resolvents of the indicated parent clauses from node N' to the final segment. It may well be impossible to perform a full relabeling in this manner because at some node the primary canceled literal is not present in the relabeled primary parent clause. The relabeling then halts and the tree structure left as is. In this case the tree is not a deduction tree.

To illustrate the above definition with Figure 2, we let the branch specified by $\{-P,Q\}$ be the primary branch and remove the node with canceled atom P. The result is a tree such as given in Figure 1 where B is $\{-P,Q\}$, C is $\{-Q\}$, A is Q and D is $\{-P\}$.

Let N be a node with canceled atom A. A <u>positive</u> (respectively, <u>negative</u>) <u>parent clause at</u> N is a parent clause containing literal A (respectively, -A) and not containing literal -A (respectively, A). Clearly, a node need not possess a positive, or negative, parent clause. If node N has a positive (respectively, negative) parent clause, the <u>positive</u> (respectively, <u>negative</u>) <u>subtree for</u> N is the deduction tree of the positive (respectively, negative) parent clause at N.

<u>Lemma 1</u>. Let N be a node with canceled atom A in a minimal refutation tree of S. Let N have a positive and a negative parent clause. Then the positive (respectively, negative) subtree for N has no clause containing literal -A (respectively, A).

<u>Proof</u>. We give the proof for the positive subtree; the other case follows analogously. Let B denote the positive parent clause at N. Suppose literal -A appears in the positive subtree for N. Then it must appear in some clause $S_1 \in S$ where S_1 appears in the positive subtree for N at an initial segment. S_1 determines (temporarily) a primary branch of the positive subtree for N. There must be a primary node N* in the positive subtree for N with canceled atom A for otherwise, by the Fact stated earlier, B would contain literal -A. Therefore, the secondary parent clause F at N* must contain the literal A. The deduction tree of F must then have an initial segment labeled by $S_2 \in S$ with $A \in S_2$. We now choose the branch from S_2 of the refutation tree as the

primary branch of the refutation tree. This branch contains nodes N* and N as primary nodes and hence specifies primary branches for the deduction tree of F and also the positive subtree for N. Now remove node N*. Further, remove any following primary node(s) which prevents the relabeling of the primary branch from being completed because of a missing primary canceled literal. The result of this modification is a smaller deduction tree. If it is a refutation tree, the original tree is not minimal, contradicting the hypothesis. Hence literal -A cannot appear in any clause of the positive subtree for N and the lemma will be proved. We now show that the resulting tree is a refutation tree.

Because of the removal of node N*, the "new" primary clauses following F in the new deduction tree may contain A, where their counterparts in the given refutation tree may not. However, in the given refutation tree clause B contains literal A and, indeed, A is the primary canceled literal at N, where N follows N* on the primary branch. Node N, which appears in the new deduction tree unless the primary parent clause does not contain A, will remove A. This assures us that the literal A, though retained in primary clauses of the new deduction tree longer than for the original refutation tree, is eliminated not to appear in the final (primary) clause of the new deduction tree. Other than this addition of a literal, each new primary clause is a (perhaps proper) subset of its counterpart primary clause in the given refutation tree. (Recall in this regard that removal of any node other than node N* occurs only when the primary canceled literal is "already" missing from the primary clause. Literals appearing in a primary clause of the given refutation tree may be missing in the counterpart new primary clause, of course, because they were introduced by a secondary parent clause of a node deleted in the new deduction tree.) But the empty clause is the only subset of itself so the final primary clause of the new deduction tree must be the empty clause. Thus the new deduction tree is a refutation tree. The lemma is proved.

<u>Corollary 1</u>. A minimal refutation tree of S contains no tautologies.

<u>Proof</u>. Suppose the tree possesses a tautology B with complementary literals A and -A. Choose as primary branch some branch containing B. Consider the last primary node N which has canceled atom A. Because the

tree is a refutation tree, N must have a positive and a negative parent clause, for otherwise the extra literal with atom A in one of the parent clauses must be present in the resolvent and hence, by Fact, in the empty clause. Contradiction. But then Lemma 1 is valid at node N. But B must be in either the positive or the negative subtree for N so cannot contain both A and -A.

<u>Remark</u>. By Corollary 1, in a minimal refutation tree every node has a positive and a negative parent clause.

<u>Corollary 2</u>. If two nodes lie on the same branch of a minimal refutation tree, then they do not have the same canceled atoms.

<u>Proof</u>. Suppose nodes N_1 and N_2 both have canceled atom A and suppose N_1 precedes N_2 on some branch of the refutation tree. By the preceding remark, N_2 has a positive and negative parent clause, hence a positive and negative subtree. N_1 must be in either the positive or negative subtree for N_2. But then either the primary or secondary canceled literal at N_1 must be missing by definition of positive (negative) subtree. Contradiction. The corollary is proved.

Suppose we are given a refutation tree of S with a designated primary branch and primary node N. We say a set J of literals <u>satisfies the * condition</u> (at N) if every literal in J is the primary canceled literal of a node following node N on the primary branch.

<u>Lemma 2</u>. Given a minimal refutation tree of S with a designated primary branch and designated primary node N, if D denotes the resolvent of primary parent clause B and secondary parent clause C, if B' is a subset of B containing the primary canceled literal and if J is a set of literals satisfying the * condition and disjoint from B', then there exists an s-linear deduction of a set $J \cup D'$ from $\{J \cup B'\} \cup S$, where $D' \subseteq D$. Moreover, the clause $J \cup B'$ need appear only as the first near parent clause of the s-linear deduction.

Before giving the proof of Lemma 2, let us see how it yields a proof of the Theorem.

Proof of Theorem (assuming Lemma 2). From remarks made earlier in the paper, we recall it suffices to prove the existence of a (ground) s-linear refutation from the existence of a (ground) refutation of S. Clearly, a minimal refutation tree of S exists if a refutation of S exists, so we may assume the given refutation tree of S is minimal. We are free to choose any branch as primary branch; we may base our selection on which clause we wish as the first near parent clause of our s-linear deduction. Our choice for first near parent clause must be an initial clause of some minimal refutation tree. The choice determines the primary branch. (This freedom allows us to assert about the general procedure that if a clause of S has an instance in a minimal refutation tree of S then there exists an s-linear refutation of S with the clause as the first near parent clause). We assume now a primary branch has been selected.

Let $E_1, E_2, \ldots, E_n$ be the sequence of clauses of the primary branch. In particular, $E_1 \in S$ and $E_n = \Box$. The s-linear deduction we now define has E_1 as the first near parent clause. A sequence of the members of S which appear on the refutation tree of S (with E_1 last) forms the prefix of the deduction. It suffices to show for $i=1,2,\ldots,n-1$ how to obtain an s-linear deduction of some set E'_{i+1}, where $E'_{i+1} \subseteq E_{i+1}$, from $S \cup \{E_i'\}$ where $E_i' \subseteq E_i$ if we demand the s-linear deduction contain E_i' only as the first near parent clause. The juxtaposition of these deductions (with prefixes removed) appended to the above-mentioned prefix forms the desired s-linear deduction.

If $E_i' \subseteq E_{i+1}$, let $E'_{i+1} = E_i'$ and the required s-linear deduction is the empty sequence. If $E_i' \nsubseteq E_{i+1}$, it must be because E_i' contains the primary canceled literal of node N separating E_i from E_{i+1} in the refutation tree. But then we apply Lemma 2 with J taken as the empty set. This yields immediately the set E_{i+1}' and the (existence of the) required s-linear deduction. The theorem is proved.

We now give the proof of Lemma 2.

Proof of Lemma 2. The proof is by induction on the size n of the secondary subtree at N. Size of a subtree is measured by the number of directed line segments (or number of clauses counting duplications) in the subtree.

Case n=1. The secondary clause C is a clause in S as it must label an initial segment of the refutation tree. The resolvent of $J \cup B'$ and C is of form $J \cup D'$ where $D' \subseteq D$ and J and D' are disjoint. The desired s-linear deduction is the sequence C, $J \cup B'$, $J \cup D'$. We must show that none of these clauses is a tautology. By Corollary 1, C is not a tautology. Let L be a literal of J with atom A. Let N' be the last primary node with canceled atom A. Such a node exists as J satisfies the * condition. J cannot also contain complementary literal $\overline{L}$ for then it is a primary canceled literal at a node N" which must precede N' on the primary branch. But both N' and N" have the same canceled atom, violating Corollary 2. Thus J is not a tautology. Also clauses B and D are in the same subtree of node N' as the primary parent clause of N' which contains literal L. Thus neither clauses B or D contain $\overline{L}$ so neither $J \cup B'$ or $J \cup D'$ is a tautology. (Recall we know B and D are not tautologies by Corollary 1).

Case n=k, assuming the result true for n < k. Let L denote the primary canceled literal at node N. Because C contains $\overline{L}$ there is a clause $S_1 \in S$ within the deduction tree of C such that $\overline{L} \in S_1$. $J \cup (B'-\{L\}) \cup S_1'$, with $J \cup (B'-\{L\})$ and S_1' disjoint, is the resolvent of $J \cup B'$ and S_1. Here $S_1' \subseteq S_1$. The s-linear deduction begins with $B_1, B_2 \ldots, B_m$, $J \cup B'$, $J \cup (B'-\{L\}) \cup S_1'$ where $B_1, \ldots, B_m$ lists the members of S. These clauses are shown to be non-tautologous in the same manner as the clauses in Case 1. $J \cup B'$ is the first near parent clause of the s-linear deduction.

It is convenient to represent these clauses in a different notation. Define J* as the set $J \cup B'-\{L\}$. Then we may write $J \cup (B'-\{L\}) \cup S_1'$ as $J^* \cup E_1'$ where $S_1' = E_1' \subseteq E_1 = S_1$ (so E_1' and J* are disjoint). Thus the first two near parent clauses of the s-linear deduction desired are $J^* \cup \{L\}$ and $J^* \cup E_1'$. We now choose a new primary branch for the refutation tree, namely, that branch which begins with S_1. Note that the branch passes through node N but that C is now the primary parent clause and B the secondary parent clause at N. All terms hereafter refer to this new choice of primary branch. We let the sequence $E_1, E_2, \ldots, E_m$ denote the primary clauses through C, e.g. $E_1 = S_1$ and $E_m = C$. The primary clauses after E_m were the primary clauses following B under the choice of primary branch given by statement of the Lemma. All the literals of $B-\{L\}$ hence are primary canceled literals of nodes following N in the new primary branch as well as in the "old" primary branch. Hence J* satisfies the * condition with the new primary branch at any node preceding and including node N. We develop the s-linear deduction sequence after $J^* \cup E_1'$ to $J^* \cup E_m'$ in the same manner as we proved the Theorem using this Lemma. Note that the secondary deduction trees at the nodes preceding N are smaller than the deduction tree for C so the induction hypothesis may be invoked to use the Lemma. We recall the manner of obtaining an s-linear deduction of $J^* \cup E_{i+1}'$, for a suitable E_{i+1}', from $\{J^* \cup E_i'\} \cup S$ for $i=1,2,\ldots,m-1$. If $E_i' \subseteq E_{i+1}$, let $E_{i+1}'=E_i'$ and the desired deduction is the empty sequence. Otherwise, E_i' contains the primary canceled literal of the node N' between E_i and E_{i+1} so by induction hypothesis we have a clause $E_{i+1}' \subseteq E_{i+1}$, which we may also assume is disjoint from J^*, and an s-linear deduction of $J^* \cup E_{i+1}'$ with $J^* \cup E_i'$ as first near parent clause. Each of these deductions (minus their prefixes) for $i=1,2,\ldots,m-1$ are fitted together in sequence and appended to the beginning sequence of clauses named above to give an s-linear deduction of $J^* \cup C'$ from $\{J \cup B'\} \cup S$. The Lemma assures us no tautologies appear in the deduction. If $\bar{L} \notin C'$ then $J^* \cup C'$ may be written as $J \cup D'$ for a $D' \subseteq D$ with D' disjoint from J because $C' \cup B'-\{L\} \subseteq D$. However, $\bar{L}$ may appear in C'. In this case, we use the subsumption option of condition (ii) of the definition of an s-linear deduction. We resolve $J^* \cup \{L\}$ with $J^* \cup C'$ to obtain $J^* \cup C'-\{\bar{L}\}$ which meets the condition that the resolvent subsume its near parent clause. $J^* \cup C'-\{\bar{L}\}$, which may be written as $J \cup D'$ for a suitable D' as above, becomes the final clause in the s-linear deduction. This clause is certainly not a tautology if its predecessor is not. The Lemma is proved.

Suppose we remove from the definition of s-linear deduction the requirement that no tautology appear in the deduction. Then Lemma 2 can be proved as stated except that "a minimum refutation tree" may be replaced by simply "any refutation tree". The proof is as given with the sections concerning tautologies removed. The "practical" significance is that by making less strict the requirements for an acceptable deduction, one does obtain refutations "beginning with" (i.e. having as first near parent clause) members of S for which no true s-linear deduction exists. Indeed, by allowing tautologies, one may begin with any clause of S which appears in some refutation tree of S. A simple example shows that we cannot disallow tautologies and still maintain this freedom of choice of members of S for first near parent clause.

Let S = $\{\{P,Q\},\{-P,-Q\},\{P\},\{Q\}\}$. No s-linear refutation exists with $\{P,Q\}$ as first near parent clause although such a refutation exists if tautologies are allowed.

Finally, we note that from the Theorem (and its manner of proof) the completeness of the set of support strategy of Wos, Robinson, Carson [1965] is obtained. A refutation is a refutation of S <u>with set of support</u> $T \subseteq S$ if every clause of the refutation of S which is a resolvent has at least one parent clause either a resolvent itself or a member of T.

<u>Corollary</u> (Wos, Robinson, Carson). If S is a finite unsatisfiable set of clauses and if $T \subseteq S$ is chosen such that S - T is satisfiable, then there is a refutation of S with set of support T.

<u>Proof</u>. There must exist a (ground) minimal refutation tree of a finite set of ground instances of S with an occurrence of some $T_1 \in T$ as a label for some initial segment of the refutation tree. This is true because the set of ground instances of S-T is a satisfiable set. As we noted in the proof of the Theorem from Lemma 2, it follows from the proof of the Theorem that there exists an s-linear refutation of S with T_1 as first near parent clause. The first resolvent of this s-linear deduction has T_1 as one parent clause; all other resolvents have resolvents as one parent clause. The Corollary follows.

The author would like to thank Peter Andrews, whose comments have led to corrections of several shortcomings of the original paper.

BIBLIOGRAPHY

[1968] Andrews, P. B. "Resolution with merging", J.ACM, 15,3 (July 1968), 367-381.

[1965a] Robinson, J. A. "A machine-oriented logic based on the resolution principle," J.ACM, 12, 1 (Jan. 1965), 23-41.

[1965b] Robinson, J. A. "Automatic deduction with hyper-resolution", Int. J. Computer Math. 1 (1965), 227-234.

[1965] Wos, L., G. A. Robinson and D. F. Carson. "Efficiency and completeness of the set of support strategy in theorem proving", J.ACM 12, 4 (Oct. 1965), 536-541.

REFINEMENT THEOREMS IN RESOLUTION THEORY*

David Luckham

ABSTRACT: The paper discusses some basic refinements of the Resolution Principle which are intended to improve the speed and efficiency of theorem-proving programs based on this rule of inference. It is proved that two of the refinements preserve the logical completeness of the proof procedure when used separately, but not when used in conjunction. The results of some preliminary experiments with the refinements are given.

*Presented at the IRIA symposium on Automatic Deduction, Versailles, France, December, 16-21, 1968.

The research reported here was supported in part by the Advanced Research Projects Agency of the Office of the Secretary of Defense (SD-183).

§1 Introduction

In practical experiments with automatic deduction programs based on the Resolution Principle of J.A. Robinson, it has been necessary to restrict the deductions (resolvents) generated by the program to some subset of all the deductions that can be made from a given set of initial hypotheses. All but the simplest of the now standard exercises for these programs would be beyond their capacity unless such restrictions were made. This is so simply because the memory space limitation would be exceeded before a proof was found. In fact, the value of experimental results such as those reported in [3] or [8] lies not so much in showing that proofs of basic theorems in certain elementary theories can be obtained (this was already established in [7]) as in gaining information about how the methods of restriction help (or do not help). Perhaps it is worth mentioning here that these studies should be viewed as part of an overall plan of working towards constructing on-line interactive deduction programs; programs that will provide a basis for systems for question-answering, proof-checking, and so on. For applications like this, it is probably not necessary to be able to prove "in one bite" so to speak, theorems an order of difficulty beyond what can now be done. Of course, it would be nice if one could! But it is quite clear that we must learn as much as possible about basic proof procedures for first order logic.

Some of the most useful methods for restricting the deductions operate by providing a condition on finite sets of clauses so that the program generates deductions from only those sets satisfying the given condition. Below we shall discuss some of those restrictions that may be characterized as follows[1]. Let $R^n(S)$ denote the set of all resolvents of level $1 \leqq n$ deducible from the initial set S, of clauses, let $R(A,B)$ denote the set of resolvents of clauses A and B, let $P(A,B)$ be a condition on pairs of clauses, and let $\tilde{R}^n(S)$ denote the subset of $R^n(S)$ defined by:

$$\tilde{R}^0(S) = S, \quad \tilde{R}^{n+1}(S) = \{C \mid (C\in R(A,B) \ \& \ A,B\in \tilde{R}^n(S) \ \& \ P(A,B)) \vee C\in \tilde{R}^n(S)\}.$$

It turns out that these methods often yield a refinement of the Resolution Principle in the sense that $\tilde{R}^n(S)$ is a proper subset of $R^n(S)$ for all n, or for all $n \leq n_0$, (or even stronger, $R^n(S) \not\subseteq \tilde{R}^m(S)$ for any $m,n > 0$), and at the same time the completeness of the proof procedure is preserved; a refinement theorem is simply a completeness theorem for such a restriction.

Although it was originally the memory space problem that motivated the development of the first refinements, it is becoming increasingly clear from experiments with more sophisticated strategies that we have reached a stage where the computation time is now an equally important problem. This is especially evident if one has on-line interactive applications in mind. Indeed, it is quite often necessary to permit the occurrence of something as bad as duplication of clauses in memory by restricting the application of time-consuming "editing" strategies such as subsumption. (This particular test, "does A subsume B", is very expensive in terms of time and it would be useful to know the most efficient way to implement it .) If the conditional test is simple, the refinement can help to reduce the pressure on both of the computation bounds, space and time, for some theorems; the extra time to perform the test will be offset by fewer editing computations. It is for this reason that we are restricting the discussion here to some very simple refining conditions on pairs of clauses only.

Section 2 contains some of the more or less standard terminology and definitions that are used in the following sections; this, it is hoped, will make the paper self-contained. Section 3 is devoted to a discussion of three refinements and the results of some preliminary experiments incorporating two of these refinements are given. Some questions concerning the use of pairs of the refinements in conjunction, remain open. Finally, Section 4 is devoted to the proofs of theorems and corollaries stated in Section 3.

§2 Notation, Terminology and Definitions

First, let us review some of the notation and terminology (of references [4, 2, 3]), which will be used in the discussion below.

The following symbols denote the concepts indicated: A,B,C,.... denote clauses; Nil denotes the empty clause; S , a finite set of clauses; H , the Herbrand domain of terms composed from the variables and function symbols in S; K, a finite subset of H; θ, λ, ρ, τ... denote substitutions (i.e., operations of simultaneously substituting a finite set of terms, t_1, ... t_n for distinct variables x_1, ..., x_n respectively); H(S), the set of all instances of S obtained by substituting terms from H (i.e., the Herbrand expansion of S); R(A,B), the finite set of resolvents of clauses A and B; R(S), the set of all clauses in S and all resolvents of pairs of clauses in S; $R^{n+1}(S) = R(R^n(S))$.

Definitions

(1) Let θ be a substitution that replaces variables by variables so that A and B θ have no variables in common. Let $\mathcal{L} \subseteq A$ and $\mathcal{M}\theta \subseteq B\theta$ be subsets of literals in A and B θ respectively with the property that there is a substitution λ such that $\mathcal{L}\lambda = \{l\}$ and $\mathcal{M}\theta\lambda = \{\neg l\}$ where l is a single literal. $\mathcal{L}$ and $\mathcal{M}\theta$ are said to be unifiable. In this case the unification algorithm [see 2 or 4] yields a unifying substitution λ_0 such that $\mathcal{L}\lambda_0 = \{l'\}$ and $\mathcal{M}\theta\lambda_0 = \{\neg l'\}$, which is "simplest" in the sense that if λ is any substitution unifying $\mathcal{L}$ and $\mathcal{M}\theta$, there is a τ such that $\lambda_0\tau = \lambda$. For $\mathcal{L}$, $\mathcal{M}$, λ_0 as above, the clause $(A-\mathcal{L})\lambda_0 \cup (B\theta - \mathcal{M}\theta)\lambda_0$ is a resolvent of A and B.

(2) If l is a literal such that $l \in A$ and $\neg l \in B$ then the clause, $(A-\{l\}) \cup (B-\{\neg l\})$ is a ground resolvent of A and B.

A ground resolvent of two clauses is a resolvent obtained by complementary literal elimination performed on the clauses themselves (not on substitution instances of the clauses); in this case we say that l is the literal resolved upon. Sometimes it is possible (and helpful) to reduce a (meta) problem to a simpler situation where one need consider only ground resolvents; we shall do this in Section 4.

A deduction or proof of a clause A from the set S (of hypotheses) is a sequence of clauses $\{D_1, D_2, \ldots, D_m\}$ such that $D_m = A$ and for all $i \leqq m$, either $D_i \in S$ or $D_i \in R(D_j D_k)$ for some $j,k < i$. A deduction may be represented in the usual way by a binary proof tree.

Essentially, the idea here is that the maximal nodes of the tree (those that do not lie below another node) are labeled by members of S ; a node which is not maximal must lie immediately below exactly two nodes and is labeled by a resolvent of the clauses at those nodes. An example is given in Figure 3. Such a proof tree may be defined as a binary, transitive, irreflexive ordering on a finite set (of nodes), together with a "labeling" function which associates a clause in $R^n(S)$ with each node, but we shall omit the details here; they may be found in [Andrews,1]. We denote the <u>proof tree</u> of A by Tr(A). Also, we shall use some of the terminology in [Andrews,1]: the minimal node of Tr(A) (labeled by A) is called the <u>root node</u>; a maximal node is a <u>leaf node</u>. The set of clauses at the leaves is the <u>base set</u> of the proof tree. We shall also need the concept of the <u>level</u> (or finite ordinal) of a proof tree:

<u>Definition</u> The <u>level</u> of Tr(A) [notation, l(Tr(A))] is defined as follows. Let u,v,w be nodes of Tr(A) and let r be the root node. Let l(u) be the function defined by: if u is a leaf node, l(u) = 0; if u is immediately below v and w, $l(u) = \max\{l(v) + 1, l(w) + 1\}$

$$l(Tr(A)) = \text{df. } l(r) .$$

Finally, for our purposes it is only necessary to consider the satisfiability of a Skolem free-variable conjunction of clauses S , over the domain H. Consequently, we adopt the following definition of <u>model</u> (or <u>interpretation</u>) and <u>satisfiability</u>.

<u>Definitions</u> (1) M is a <u>model</u> (or <u>interpretation</u>) = df. M is a non-empty set of literals and negations of literals occurring in H(S) such that for all literals l in H(S) ,

$$l \in M \Leftrightarrow \neg l \notin M$$

(2) $\overline{M} =$ df. $\{l \mid \neg l \in M\}$

(3) M <u>satisfies</u> a clause C (notation, $M \models C$) = df.

$M \cap C_i \neq \emptyset$ for all instances $C_i \in H(C)$.

(4) M is a <u>model for S</u> (or M satisfies S ; notation, $M \models S$) = df. M satisfies each clause in S .

Thus free variables are interpreted as though they were universally quantified, and a set S is interpreted as the conjunction of its members. The conjunction of all the instances of clauses in H(S) is truthfunctionally inconsistent if and only if S is not satisfiable.

3 Refinements of the Resolution Principle

Let $\tilde{R}_1$, $\tilde{R}_2$ and $\tilde{R}_3$ denote the following refinements:

(1) Resolution relative to a model M:

Definition $\tilde{R}_1^{\,0}(S) = S$

$$\tilde{R}_1^{\,n+1}(S) = \{C \mid C \in R(A,B) \ \& \ A,B \in \tilde{R}_1^{\,n}(S)$$

$$\& \ (M \nvDash A \vee M \nvDash B)\} \cup \tilde{R}_1^{\,n}(S).$$

In resolution relative to a model M , only those resolvents following from at least one clause not satisfied by M , are generated. This idea is a direct generalization of the notion of P_1 - deduction presented in [5]. In practice, given S , the choice of M is of course crucial in determining whether the resulting refinement is of any value.

Theorem 1 If S is unsatisfiable there is an m such that $\text{Nil} \in \tilde{R}_1^{\,m}(S)$.

Thus $\tilde{R}_1$ is complete. This theorem was obtained independently by the author and J. Slagle [6] who has studied more complex refinements; the proof given here (in Section 4) makes use of a form of the "Maximal Model" lemma of G. Robinson and L. Wos [9]. Some additional information about $\tilde{R}_1$ may be extracted from the proof. First, the terms in a proof tree, $Tr_1(\text{Nil})$, occurring, say in $\tilde{R}_1^{\,m}(S)$ need be no more complex than the terms in any proof tree, Tr(Nil) occurring in $R^n(S)$. This means that when using resolution relative to a model it is not necessary to make an adjustment to the standard editing strategy of limiting the depth of nesting of function symbols in the terms. The same is not true however of the strategy which limits the length of clauses. It is therefore, probably better to use separate editing strategies of this sort, rather than some "combination" strategy such as bounding the total number of symbols in a clause. Secondly, if there is a proof in $\tilde{R}_1^{\,m}(S)$, there is one without tautologies (corollary 2). It is also possible to derive the completeness of the set of support strategy [7] in a strong form, namely that it is unnecessary to compute resolvents of pairs of clauses in S-T even when T-supported derivations of these clauses occur. Finally, the completeness of the model partition strategies of [3] is a consequence of theorem 1.

The computational complexity of the conditional test for generating resolvents relative to a model depends on the recursion equations defining the model. So far, only very simple models have been used in practical experiments and the results are encouraging (for a discussion see [3]).

(2) Resolution with Merging (P. Andrews[1]):

Definition Let A be a resolvent of B and C :

$A = (B-\mathcal{L})\rho \cup (C\theta - \mathcal{M}\theta)\rho$. A is a merge if

$(B-\mathcal{L})\rho \cap (C\theta - \mathcal{M}\theta)\rho \neq \emptyset$. Similarly, a substitution instance

$A\lambda$ is a merge instance if $(B-\mathcal{L})\rho\lambda \cap (C\theta - \mathcal{M}\theta)\rho\lambda \neq \emptyset$.

The property of being a merge is therefore a property of a particular occurrence of a clause A in $R^n(S)$ and depends on the proof tree for that occurrence. An occurrence of a resolvent which is not a merge is called a non merge (occurrence). A clause may occur both as a merge and as a non merge, and this raises certain implementation problems if one wants to make the most of the refinement strategy proposed below.

Definition $\tilde{R}_2^{\,0}(S) = S$

$\tilde{R}_2^{\,n+1}(S) = \{C \mid C \in R(A\lambda, B\tau)$ & $A, B \in \tilde{R}_2^{\,n}(S)$ & $(A \in S \vee B \in S \vee$

$A\lambda$ is merge $\vee$ $B\tau$ is a merge$)\} \cup \tilde{R}_2^{\,n}(S)$.

Essentially, Andrews has proved in [1] that, starting with base set S , if there is a proof tree Tr(Nil), then there is a proof tree Tr'(Nil) in which no clause occurs as the resolvent of two non merge instances; he also shows that Tr'(Nil) satisfies the set of support condition. Thus $\tilde{R}_2$ is complete with the set of support strategy.

iii) Resolution relative to the Ancestry Filter:

efinition $\tilde{R}_3^{\,0}(S) = S$

$$\tilde{R}_3^{\,n+1}(S) = \{C \ : \ C \in R(A,B) \ \& \ A,B \in \tilde{R}_3^{\,n}(S) \ \&$$

$$(A \in S \vee B \in S \vee A \in Tr(B) \vee$$

$$B \in Tr(A))\} \cup \tilde{R}_3^{\,n}(S).$$

At each step in the generation of $\tilde{R}_3^{n+1}(S)$ a clause A is resolved ith those clauses B in $\tilde{R}_3^{\,n}(S)$ satisfying the condition, $B \in S \vee B \in Tr(A)$. he proof tree of A itself provides the restriction or filter on $\tilde{R}_3^{n}(S)$. It is asy to see that the deduction sequence of $Tr(A)$ may be written (if we allow epetitions) in the form $\{C_1 C_2 D_1 B_1 D_2 \dots D_i B_i D_{i+1} \dots A\}$ where $C_1, C_2 \in S$, $B_i \in S$ r $B_i = D_j$, for some $j < i$, and $D_{i+1} \in R(D_i, B_i)$ (see Figure 3). We shall say hat such proof trees are in Ancestry Filter Form (notation, AFF) relative to the ase set S, and we choose to call C_1 the top node of the tree; each resolvent, $_1$, D_2, ..., A is below C_1. The branch from C_1 to A is called the principle ranch of the tree.

Notice that if $B_i = D_j$ then the sub-tree $Tr(B_i)$ is identical with he initial segment $Tr(D_j)$ in the sense that the trees are not only isomorphic but lso the clauses at corresponding nodes are the same (e.g. $Tr(23)$ and $Tr(2)$ in igure 3).

There are two points that are perhaps worth mentioning here. First, the estriction of $\tilde{R}_3$ does not come into force until level 2; therefore it is mportant to show that $\tilde{R}_3$ may be used in conjunction with some other refinement f level 1 such as the Set of Support. Secondly, it would be nice if the refining ondition of $\tilde{R}_3$ could be replaced by $(A \in S \ \vee \ B \in S)$. Although most of the heorems proved (by programs) so far have proofs satisfying this condition, in eneral it yields an incomplete procedure (as is pointed out in Andrews [1]).

<u>Theorem 2</u> $\tilde{R}_3$ is complete, and it is complete with the Set of Support strategy.

One of the problems involved in applying refinements of the sort suggested by Theorems 1 and 2 is that the level of the proof tree obtained is generally greater than the shortest possible proof. One is therefore faced with the "trade-off" of generating fewer clauses at each level against searching deeper levels for a proof. As has been shown in [3], it is hard to find any relationship between the level of the proof tree and the efficiency (in terms of the percentage of deductions retained which did <u>not</u> contribute to the proof) with which the proof was found. In general, the refinements of Theorem 1 are certainly worthwhile using to find proofs of simple basic theorems in algebra and number theory, and will probably be more useful as the theorems become more difficult.

Some preliminary comparisons have been made between resolution relative to the ancestry filter with Set of Support, T, with Unit Preference, against resolution relative to a model for S-T and Unit Preference. These are given in Figure 1 and tend to support the plausible view that the ancestry filter will do better on more difficult theorems whose proofs depend on lemmas. It is also interesting to compare the refinement strategies with the combination of Set of Support and Unit Preference strategies that forms a very powerful part of the theorem-proving programs discussed in [7,8]. The elementary number theory proof in Example 5 illustrates two features very clearly. The proof given in Figure 2 and 3 was obtained with a conjunction of the refinements $\tilde{R}_1$ and $\tilde{R}_3$ and is a level 12 proof; the combination of Set of Support (corresponding to the model used in $\tilde{R}_1$) and unit preference generates a level 7 proof. However, the latter combination places such a weight on the editing strategies that the level 12 $\tilde{R}_1 \cap \tilde{R}_3$ proof was obtained in half the time (we note that the program is written in LISP; it would not be hard to improve these times by a factor of 10).

It is natural to try using the refinements in conjunction, as was done in Example 5. $\tilde{R}_3$ is complete with the Set of Support; it might be hoped that it would be complete with $\tilde{R}_1$ (which may be viewed as a generalization of Set of Support). Unfortunately, this is not the case. (We denote the conjunction of

$\tilde{R}_i$ and $\tilde{R}_j$ by $\tilde{R}_i \cap \tilde{R}_j$):

Theorem 3 (i) $\tilde{R}_1 \cap \tilde{R}_2$ is incomplete.

(ii) $\tilde{R}_1 \cap \tilde{R}_3$ is incomplete.

Proof: Consider the set of clauses, $C_1 = \{\neg r\}$, $C_2 = \{r, \neg q\}$, $C_3 = \{\neg p, r\}$, $C_4 = \{p, q\}$, together with the model $M = \{p, q, r\}$. Let $\tilde{R}_1$ be resolution relative to M. It is easy to see that although $\{C_1, C_2, C_3, C_4\}$ is a (minimally) inconsistent set, there is no proof tree of Nil satisfying $\tilde{R}_1 \cap \tilde{R}_2$ or $\tilde{R}_1 \cap \tilde{R}_3$.

Q.E.D.

At the moment we do not know if the conjunction $\tilde{R}_2 \cap \tilde{R}_3$ is a complete refinement[3]. Another interesting question (in view of applications such as Example 5) is whether there exists a decision procedure (and indeed a practicable one) for recognizing, given S and M, whether or not there is a proof tree Tr(Nil) with base set in S satisfying $\tilde{R}_1 \cap \tilde{R}_3$. If the answer to this question is no, there ought at least to be some useful sufficient conditions for recognizing incomplete situations.

<u>Example 1</u>: In a closed associative system with left and right solutions to equations, there is an identity element.

<u>Example 2</u>: In a group the right identity is also a left identity.

<u>Example 3</u>: In a group with right inverses and right identity, every element has a left inverse.

<u>Example 4</u>: If an associative system has an identity element and the square of every element is the identity, then the system is commutative.

	$\tilde{R}_3$		$\tilde{R}_1$	
	No. of clauses retained	level	No. of clauses retained	level
Ex.1	66	4	44	4
Ex.2	34	6	36	6
Ex.3	94	7	96	7
Ex.4	60	10	94	10

Figure 1

Example 5   If a is a prime number and $a = {}^{b^2}/c^2$ then a divides both b and c.

In the hypotheses below[2], interpret P(x) as "x is a prime", M(x,y,z) as "x X y = z", D(x,y) as "x divides y", S(x) as "x^2", and F(x,y) as the number of times x divides y. Hypothesis 6 states that if a prime divides the product of two integers, it divides at least one of the integers; hypothesis 7 is a consequence of the cancellation law.

With hypotheses {1,2,3} as a set of support and unit preference with level bound 5, a proof was obtained at level 7, clauses retained = 59, clauses generated = 144, time = 75 seconds.

With resolution relative to the ancestry filter and the model M containing all negated literals, but no unit preference, the proof below (and Figure 3) was obtained at level 12, clauses retained = 66, clauses generated = 73, time = 37 seconds.

Hypotheses:

1. P(A)
2. M(A, S(C), S(B))
3. M(X1,X1, S(X1))
4. M(X2,X1,X3) ¬M(X1,X2,X3)
5. D(X1,X3) ¬M(X1,X2,X3)
6. D(X1,X3) D(X1,X2) ¬P(X1) ¬M(X2,X3,X4) ¬D(X1,X4)
7. M(X1, S(F(X1,X2)), S(X3)) ¬M(X1,S(X3), S(X2)) ¬D(X1,X2)
8. ¬D(X1,C) ¬D(X1,B).

Proof   ($\bar{R}_1 \cap \bar{R}_3$):

NIL 1 2

1 ¬D(A,B) 3 4
2 D(A,B) 5 6
3 D(A,C) 7 8
4 ¬D(X1,C) ¬D(X1,B) AXIOM 8

(Figure 2 continued on next page)

```
5  D(A,X3) D(A,X2) ¬M(X2,X3,S(B))  9 10
6  M(X1,X1,S(X1)) AXIOM 3
7  D(A,X3) D(A,X2) ¬M(X2,X3,S(C))  11 12
8  M(X1,X1,S(X1)) AXIOM 3
9  D(A,X3) D(A,X2) ¬P(A) ¬M(X2,X3,S(B))  13 14
10 P(A) AXIOM 1
11 D(A,X3) D(A,X2) ¬P(A) ¬M(X2,X3,S(C))  15 16
12 P(A) AXIOM 1
13 D(A,S(B))  17 18
14 D(X1,X3) D(X1,X2) ¬P(X1) ¬M(X2,X3,X4) ¬D(X1,X4) AXIOM 6
15 D(A,S(C))  19 20
16 D(X1,X3) D(X1,X2) ¬P(X1) ¬M(X2,X3,X4) ¬D(X1,X4) AXIOM 6
17 D(X1,X3) ¬M(X1,X2,X3) AXIOM 5
18 M(A,S(C),S(B)) AXIOM 2
19 M(A,S(F(A,B)),S(C))  21 22
20 D(X1,X3) ¬M(X1,X2,X3) AXIOM 5
21 M(A,S(F(A,B)),S(X3)) ¬M(A,S(X3),S(B))  23 24
22 M(A,S(C),S(B)) AXIOM 2
23 D(A,B)  25 26
24 M(X1,S(F(X1,X2)),S(X3)) ¬M(X1,S(X3),S(X2)) ¬D(X1,X2) AXIOM 7
25 D(A,X3) D(A,X2) ¬M(X2,X3,S(B))  27 28
26 M(X1,X1,S(X1)) AXIOM 3
27 D(A,X3) D(A,X2) ¬P(A) ¬M(X2,X3,S(B))  29 30
28 P(A) AXIOM 1
29 D(A,S(B))  31 32
30 D(X1,X3) D(X1,X2) ¬P(X1) ¬M(X2,X3,X4) ¬D(X1,X4) AXIOM 6
31 D(X1,X3) ¬M(X1,X2,X3) AXIOM 5
32 M(A,S(C),S(B)) AXIOM 2
QED
```

Figure 2

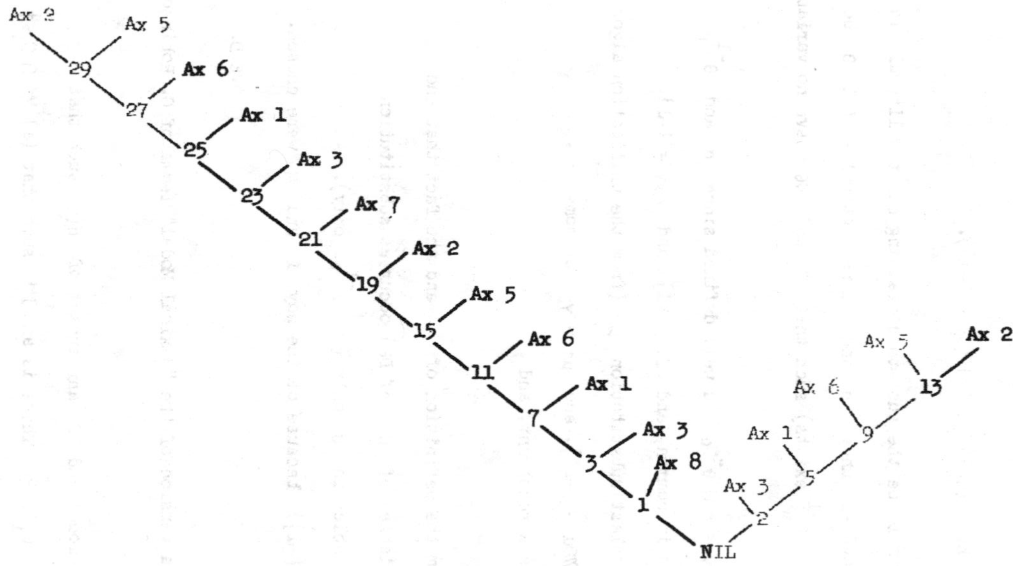

Ancestry filter form proof tree for the proof in Example 5; notice Tr(23) is a renumbering of Tr(2).

Figure 3

§4 Proof of Theorems

<u>Lemma 1</u> Let $C\sigma$ and $D\rho$ be instances of clauses C and D containing the literals l and $\neg l$ respectively. Then there is a resolvent E of C and D and a substitution τ such that

$$E\tau = (C\sigma - \{l\}) \cup (D\rho - \{\neg l\}).$$

<u>Proof</u> Let $\mathcal{L} \subset C$ and $\mathcal{M} \subset D$ be the subsets containing all the literals in C and D that unify to l and $\neg l$ under σ and ρ respectively. Let θ be a (1-1) change of variables (so θ^{-1} exists) such that C and $D\theta$ have no variables in common.

The substitution $\gamma = \sigma \cup \theta^{-1}\rho$ is well defined since σ and $\theta^{-1}\rho$ have no substitution variables in common, and $\mathcal{L}\gamma = \{l\}$ and $\mathcal{M}\theta\gamma = \{\neg l\}$.

Thus there is a simplest substitution μ (from the unification algorithm) such that $\mathcal{L}\mu = \{l'\}$, $\mathcal{M}\theta\mu = \{\neg l'\}$, and $\mu\tau = \gamma$ for some τ.

Then $E = (C-\mathcal{L})\mu \cup (D-\mathcal{M})\theta\mu$ is a resolvent, and

$E\tau = (C-\mathcal{L})\gamma \cup (D-\mathcal{M})\theta\gamma$

$= (C-\mathcal{L})\sigma \cup (D-\mathcal{M})\rho$ from the definition of γ and the fact that the variables of C (of $D\theta$) occur as substitution variables in σ only (in $\theta^{-1}\rho$ only).

$= (C\sigma - \{l\}) \cup (D\rho - \{\neg l\})$ because of the way $\mathcal{L}$ and $\mathcal{M}$ were chosen.

Q.E.D.

The next lemma is a version of the "Maximal Model" lemma of G. Robinson and L. Wos [9]:

<u>Lemma 2</u> Let M be a given model and P any subset of $H(S)$ such that $M \cap C_i \neq \emptyset$ for all instances $C_i \in P$. There is an M^* such that (i) $M^* \cap C_i \neq \emptyset$ for all $C_i \in P$, and (ii) $M^* \cap \overline{M}$ is maximal under condition (i). In particular, for any literal $l \in M^* \cap M$, $\neg l$ is in $\overline{M} - M^*$; hence there is an instance $C_i \in P$ such that $M^* \cap C_i = \{l\}$.

Proof. Let $\mathcal{U}$ be any subset of $H(S)$ and let $\mathcal{U}^c$ denote the closure of $\mathcal{U}$ under complementary literal elimination (c.l.e.).

First we note that M^* satisfies the lemma with respect to P if and only if it satisfies the lemma with respect to P^c. For, if $M \cap C_i \neq \emptyset$ for all $C_i \in P$ then $M \cap C_k \neq \emptyset$ for all $C_k \in P^c$ (by soundness of c.l.e), and conversely (since $P \subseteq P^c$). Thus it is easily seen that (a). (i). $M^* \cap C_i \neq \emptyset$ for all $C_i \in P$, and (ii). $M^* \cap \overline{M}$ is maximal under condition (i), if and only if (b). (i). $M^* \cap C_j \neq \emptyset$ for all $C_j \in P^c$, and (ii). $M^* \cap \overline{M}$ is maximal under condition (i).

If (a) is true, then (b) (i) is satisfied; further, given any M' such that $M^* \cap \overline{M}$ is a proper subset of $M' \cap \overline{M}$, there is a $C' \in P$ such that $M' \cap C' = \emptyset$; but then $C' \in P^c$ and so (b) (ii) is also satisfied. Conversely, if (b) is true, (a) must also be true because any M' falsifying (a) (ii) would also falsify (b) (ii).

Let P^c and M be ordered according to the usual principle of increasing complexity (of terms, literals, and clauses) and for elements of a given complexity level (finite in number), according to some fixed lexical order. Thus we have two well-orderings:

$<_1$ is the ordering of P^c and $<_2$ is the ordering of M.

Let $M^*_0 = \emptyset$. For $i = 1, 2, 3, \ldots$ we construct M^*_i as follows: The negation of the ith element in $<_2$ say $\neg l_i$, is added to M^*_{i-1} to form M^*_i unless this denies a clause C_j in P^c in the sense that all the literals of C_j would then be in $\overline{M^*_i}$ (the finite set of negations of members of M^*_i). This may be decided in a finite number of steps; one simply tests each clause in $<_1$ for inclusion in $\overline{M^*_i}$ until the complexity level of a literal in a C_k is greater than the complexity level of l_i or $\neg l_i$, in which case $C \not\subseteq \overline{M^*_i}$, all clauses C such that $C_k <_1 C$, and no further tests are necessary. If a C_j is denied by this move, then $M^*_i = M^*_{i-1} \cup \{l_i\}$. Notice that in this case, $l_i \in M^*_i \cap M$ and $M^*_i \cap C_j = \{l_i\}$. Further, since P is satisfiable, an easy inductive argument shows that it is not possible for C to be denied by $M^*_{i-1} \cup \{\neg \ell_i\}$ and D to be denied by $M^*_{i-1} \cup \{\ell_i\}$; for then $E = (C - \{\neg \ell_i\}) \cup (D - \{\ell_i\})$ occurs earlier than C or D in $<_1$ and would be denied by M^*_{i-1}.

Let $M^* = \bigcup_i M_i^*$. It is easily seen that M^* satisfies the conditions of the lemma, and is recursive in M and P^c .

Q.E.D.

The following lemma may be proved by a straightforward König's lemma argument; we omit the proof.

Lemma 3 If S is unsatisfiable (i.e., has no model in our sense) there is a finite subset $K \subseteq H$ such that K(S) is inconsistent.

For the purposes of the proof below we need to introduce the concept of a set of substitution instances in which the complexity of the terms (i.e., depth of nesting of function symbols) is restricted to be no greater than the complexity of terms in K(S) . Essentially, we assume an infinite set of variables, $\overline{X} = \{x_1, x_2, x_3, \ldots\}$, containing the variables of S , and we admit any term that can be obtained from a term in K(S) by a change of variables in $\overline{X}$ (i.e., alphabetic variants from $\overline{X}$) .

Definitions (i) $K(\mathcal{U}|K)$ = df. the set of alphabetic variants of instances of clauses of $\mathcal{U}$ containing only terms in K(S) .

(ii) $R(\mathcal{U}|K)$ = df. the subset of $R(\mathcal{U})$ consisting of those clauses containing only alphabetic variants of terms occurring in K(S) .

Proof of Theorem 1

Let $\widetilde{R}$ denote resolution relative to the model M . Assume S is unsatisfiable. Hence by lemma 3 there is a finite set $K \subseteq H$ such that K(S) has no model. For any model M' there is a clause C in S and a substitution σ such that $C\sigma \in K(S)$ and $C\sigma \cap M' = \emptyset$; similarly, because $K(S) \subseteq K(\widetilde{R}^m(S)|K)$, there is a $C\sigma \in K(\widetilde{R}^m(S)|K)$ such that $C\sigma \cap M' = \emptyset$.

CLAIM: If Nil $\notin \widetilde{R}^m(S|K)$ then $\widetilde{R}^{m+1}(S|K) \neq \widetilde{R}^m(S|K)$.

In order to prove the claim, we first partition $K(\widetilde{R}^m(S)|K)$ into sets P and

$\bar{P}$:

$$P = \{C\sigma | C\sigma \in K(\tilde{R}^m(S)|K) \ \& \ C\sigma \subseteq \bar{M}\}$$

Essentially, P is the set of those instances of $K(\tilde{R}^m(S)|K)$ denied by the model M ; P cannot be empty for if it was, $K(S)$ would be consistent, contrary to our choice of K .

By lemma 2, (with $\bar{M}$ as the given model) a model M' satisfying the following two conditions exists:

(i) for all $C\sigma \in P$, $M' \cap C\sigma \neq \emptyset$

(ii) $M' \cap M$ is maximal under condition (i) .

Now we use M' to find an instance of a resolvent in $\tilde{R}^{m+1}(S|K) - \tilde{R}^m(S|K)$ as follows. First, there exists a $C\sigma \in K(\tilde{R}^m(S)|K)$ such that $C\sigma \cap M' = \emptyset$; any such $C\sigma$ must have a non-empty intersection with M since it does not belong to P ; choose $C_0\sigma_0$ so that $C_0\sigma_0 \cap M$ is minimal among such $C\sigma$.

Let $l \in C_0\sigma_0 \cap M$; then $l \in M-M'$. Therefore, $\neg l \in M' \cap \bar{M}$, and there is a $D_0\rho_0$ in P such that $M' \cap D_0\rho_0 = \{\neg l\}$.

By lemma 1 there is a resolvent E of C_0 and D_0 and a substitution τ such that

$$E\tau = (C_0\sigma_0 - \{l\}) \cup (D_0\rho_0 - \{\neg l\}) \ .$$

Since $D_0\rho_0 \subseteq \bar{M}$, the model M does not satisfy D_0 , and it follows that $E \in \tilde{R}^{m+1}(S)$, and all the terms of $E\tau$ are contained in $K(S)$ so that $E \in \tilde{R}^{m+1}(S|K)$. Notice that the computation of E from C_0 and D_0 may introduce new variables (from $\bar{X}$) not occurring in C_0 or D_0; this is why we consider alphabetic variants of terms in $K(S)$.

If $E = \text{Nil}$ we have proved the claim; otherwise it remains to show that $E \notin \tilde{R}^m(S)$. First, $E\tau \cap M' = \emptyset$, so that if $E \in \tilde{R}^m(S)$ then $E\tau \in \bar{P}$; but in this case $E\tau \cap M \subsetneqq C_0\sigma_0 \cap M$, contradicting the minimality of $C_0\sigma_0 \cap M$. This proves the claim.

Now, the process of computing sucessively $\tilde{R}^1(S|K)$, $\tilde{R}^2(S|K)$,. . . must stop after a finite number of steps since the terms are restricted to the finite depth of nesting of function symbols of terms in K(S) , which means that only a finite number of resolvents can be generated. Thus there is an m such that $Nil \in \tilde{R}^m(S|K)$. Q.E.D.

<u>Corollary 1</u> Let M be a given model. If K is a finite subset of H such that K(S) is inconsistent, then there is a proof by resolution relative to M in which only alphabetic variants of terms from K(S) occur.

<u>Corollary 2</u> If S is unsatisfiable, there is a proof by resolution relative to a given model M and the proof does not contain tautologies.

<u>Proof</u> In the proof above E_T is not a tautology since $E_T \cap M' = \emptyset$; therefore E is not a tautology.

The next corollary refers to the Set of Support Strategy; see references [7] and [8].

<u>Corollary 3</u> The set of support strategy is complete. Further, it is complete with the editing strategy of eliminating repetitions; in particular, it is unnecessary to compute resolvents of pairs of members of S-T when T-supported deductions of those members occur.

<u>Proof</u> Since S-T is satisfiable it has a model M . M cannot satisfy every clause in T unless S is consistent. Let $\tilde{R}$ denote resolution relative to this model. Then for all m the proof tree of any clause in $\tilde{R}^m(S)$ is a deduction with T-support. Furthermore since in the proof above $E \in \tilde{R}^{m+1}(S) - \tilde{R}^m(S)$, the strategy remains complete if repetitions of clauses in $\tilde{R}^m(S)$ are eliminated.

orollary 4 The model-partition strategies of [3] are complete.

roof If $\langle M_1, M_2 \rangle$ is a model partition in the sense of [3], the restricted tree obtained by applying the partition strategy is the same as that generated by resolution relative to M_2 .

We turn now to the proof of theorem 2 which relies heavily on two emmas of Andrews [1] concerning transformation of proof trees. The argument falls aturally into two parts. Part one is to show that if there is a ground resolution roof tree, Tr(A) , of a clause A with a base set of clauses in K(S) and with chosen leaf node Γ , then there is a ground resolution proof tree with the same ase set, Tr'(A'), in ancestry filter form (AFF) having Γ as top node and such hat $A' \subseteq A$.

Part two of the argument runs as follows. If there is a (general) esolution proof tree, Tr(Nil) , with base set S , then there is a ground esolution proof tree, Tr_1(Nil) with a base set in some finite set of instances, (S) (by soundness of resolution, lemma 3, and completeness of ground resolution); y part one, there is an AFF ground resolution proof tree, Tr_2(Nil), with the same ase set in K(S) ; by repeated application of lemma 1, starting at the top node of r_2(Nil) and working down the tree, there is an isomorphic general resolution tree r_3(Nil) in AFF with base set S . Completeness with the Set of Support is taken are of by using the freedom allowed in part one to choose the top node appropriate-y; the ground resolution tree Tr_1(Nil) must contain a base clause which is an nstance of the clause in the Set of Support, T (otherwise S-T is inconsistent); hoose Γ to be a leaf at which such an instance occurs; Γ becomes top node of r_2(Nil); Tr_3(Nil) then has a member of T at its top node, and is therefore a roof tree with T support. Thus, if there is a proof tree, Tr(Nil), with base et S , then there is a T supported proof tree Tr'(Nil), in ancestry filter orm with the same base set; completeness of $\tilde{R}_3$ with Set of Support now follows rom the completeness of the resolution principle which, we note, is a consequence f theorem 1.

It remains to prove the first part of the argument dealing only with ground resolution proof trees. We begin by quoting two lemmas of Andrews [1] in somewhat intuitive (but we hope, transparent) form.

First, some more terminology. (1). Given a ground resolution proof tree Tr(A), suppose that the literal l occurs in A, and that in Tr(A), $A \in R(B,C)$. Those occurrences of l in A,B or C are immediate ancestors of l in A. The transitive closure of this relationship over Tr(A) yields the set of ancestors of the particular occurrence of l in A. (2). If l' is an ancestor of l then l is a descendant of l'.

In what follows, when speaking of a node C it is convenient to mean both the node itself and the clause occurring at the node.

Definition Let Tr(A) contain a leaf node C containing a literal l. If no descendant of this (occurrence of) l is resolved upon, we say that l in C is not resolved upon in Tr(A).

Lemma 4 [Andrews, 1, lemma 3] Let Tr(A) be a ground resolution tree with base set S and let C be a leaf node containing a literal l not resolved upon Tr(A). Then

(i) A contains l, and

(ii) if in Tr(A), the leaf node C is replaced by $C' = C-\{l\}$, and all descendants of l in C having no other leaf ancestor are similarly eliminated, we obtain a proof tree Tr'(A') isomorphic to Tr(A) with base set $S \cup \{C-\{l\}\}$ and such that $A' \subseteq A$.

Lemma 5 [Andrews, 1, lemma 4] Let Tr(A) be a ground resolution proof tree with base set S. Let the leaf node C be replaced by $C \cup B$, and each node, D, below C be replaced by the appropriate $D \cup B'$ as follows: if $D = (E-\{l\} \cup (F-\{\neg l\})$ and E is replaced by $E \cup B''$, then $B' = B'' - \{l\}$. Then we obtain a ground resolution proof tree Tr'(A') isomorphic to Tr(A) having base set $S \cup \{C \cup B\}$ and such that $A' \subseteq A \cup B$.

Now part one of the argument is the next lemma:

Lemma 6 Let Tr(a) be a ground resolution proof tree with base set S, and let Γ be a chosen leaf node. Then there is a ground resolution proof tree, Tr'(A') in ancestry filter form relative to S, having Γ as top node, and such that $A' \subseteq A$

Proof: By induction on the level of Tr(A).

If $l(Tr(A)) = 1$, the lemma is trivially true. Suppose the lemma is true for trees of level n. Let Tr(A) be a proof tree of level n+1 with base set S and suppose it is not already in AFF. with the top node as the chosen leaf node Γ.

Let $A = B \cup C$ such that p is the literal resolved upon to form A, and the clauses at the two nodes immediately above A are $B_1 = B \cup \{p\}$ and $C_1 = C \cup \{\neg p\}$. Let Γ be a leaf node in $Tr(B_1)$. We note that $l(Tr(B_1)) \leq n$ and $l(Tr(C_1)) \leq n$.

Applying the induction hypothesis, there is a $Tr'(B'_1)$, with the same base set as $Tr(B_1)$, in AFF, with Γ as top node, and such that $B'_1 \subseteq B_1$. If p is not in B'_1, then $B'_1 \subseteq A$ and $Tr'(B'_1)$ satisfies the conclusion of the lemma. If $p \in B'_1$, let us write, $B'_1 = B_2 \cup \{p\}$.

In this latter case, we now consider $Tr(C_1)$. The idea is to transform this tree to $Tr'(C'_1)$ with a base clause containing $\neg p$ at the top node, and then to append $Tr'(B'_1)$, by a resolution on p, to this node. Two things can go wrong. First the ancestry relation of $\neg p$ may not be preserved under the transformation of the induction hypothesis so that resolving on $\neg p$ at the top node may spoil a resolution further down $Tr'(C'_1)$; using lemma 4 to take $\neg p$ out of a leaf node in $Tr(C_1)$ before applying the induction hypothesis, and then using lemma 5 to put B_2 into the top node of $Tr'(C'_1)$ gets around that problem. Secondly, the clause at the tope node of $Tr'(C'_1)$ may occur at other leaves as well; we will need to append $Tr'(B'_1)$ to each of these leaves in order to maintain the "finished" tree in ancestry filter form relative to S . The details follow.

Consider $Tr(C_1)$. The occurrence of $\neg p$ in C_1 must have an ancestor at a leaf node, D_i, say. Now $\neg p$ in D_i is not resolved upon in $Tr(C_1)$. Suppose $D_i = D_i' \cup \{\neg p\}$. By lemma 4 there is a proof tree $Tr_1(C_1^1)$ with base set in $S \cup \{D_i'\}$ isomorphic to $Tr(C_1)$ and such that $C_1^1 \subseteq C_1$. The isomorphism implies that $l(Tr_1(C_1^1)) \leq n$. By applying the induction hypothesis to Tr_1 with D_i' as the chosen node, there exists a tree $Tr_2(C_2)$, with base set in $S \cup \{D_i'\}$ in AFF, having D_i' as the top node and such that $C_2 \subset C_1^1$.

The clause D_i' occurs at the top node of $Tr_2(C_2)$ and possibly at other leaf nodes as well. Suppose that we have a standard procedure for enumerating the leaves of an AFF tree; e.g. we could adopt a left-to-right convention as in Figure 3 and, starting, with the top node, always move to the right. Starting with the top node, apply the replacement procedure of lemma 5 to $Tr_2(C_2)$, replacing D_i' by $D_i' \cup B_2$, to obtain an isomorphic tree $Tr_2'(C_2 \cup B_2')$ where $B_2' \subseteq B_2$; repeat this process on Tr_2' , replacing D_i' at the next leaf node at which it occurs by $D_i' \cup B_2$. This finally yields a tree $Tr_3(C_2 \cup B_3)$ with base set in $S \cup \{D_i' \cup B_2\}$ isomorphic to $Tr_2(C_2)$, having $D_i' \cup B_2$ as top node and such that $B_3 \subseteq B_2$; clearly Tr_3 is an AFF. tree since Tr_2 is an AFF tree and the replacement procedure of lemma 5 when applied to identical sub-trees transforms them into identical trees.

Next we add a single resolvent to $Tr'(B_2 \cup \{p\})$. We resolve $B_2 \cup \{p\}$ and $D_i' \cup \{\neg p\}$ upon p to obtain $D_i' \cup B_2$. This gives us a tree $Tr(D_i' \cup B_2)$ with a base set in S (since $D_i \in S$) which is clearly an AFF. tree since Tr' is an AFF. tree and the final resolvent satisfies the AFF. condition relative to S .

Now, we append $Tr(D_i' \cup B_2)$ to each leaf node in Tr_3 at which $D_i' \cup B_2$ occurs, starting with the top node. This is done by placing $Tr(D_i' \cup B_2)$ so that its root and the leaf coincide. What we have now is a tree r$_4(C_2 \cup B_3)$ with base set in S (all leaf occurrences of $D_i' \cup B_2$ have been liminated) in AFF. with the chosen node Γ at the top (See Figure 4); notice hat replacing all leaf occurrences of the top node clause in an AFF. tree by an FF. proof tree of that clause relative to S preserves the AFF. property. Γ is op node of Tr_4 since it is top node of $Tr(D_i' \cup B_2)$.

Finally, we note that $B_3 \subseteq B_2 \subseteq A$ and $C_2 \subseteq C \cup \{\neg p\}$. Suppose $p \in C_2$. We notice that in $Tr_4(C_2 \cup B_3)$ the clause $B_2 \cup \{p\}$ is an ancestor of $_2 \cup B_3$ since $Tr'(B_2 \cup \{p\})$ is an initial segment of Tr_4 . We may therefore dd the resolvent of $C_2 \cup B_3$ and $B_2 \cup \{p\}$ upon p to Tr_4 without violating he AFF. condition (See Figure 4). This yields $Tr_5(C_2' \cup B_2)$ which satisfies the emma since $C_2' \cup B_2 \subseteq A$.

This completes the proof of lemma 6 . The proof provides a recursive rocedure for transforming a proof tree $Tr(A)$ with base set S and chosen leaf into an AFF. tree relative to S , $Tr'(A')$, in which Γ is top node and $' \subseteq A$.

Q.E.D.

cknowledgements The author wishes to thank NILS NILSSON for many stimulating iscussions on the subject matter treated here, especially theorem 2, and JOHN LLEN for programming and providing the examples 1 to 5 in Section 3.

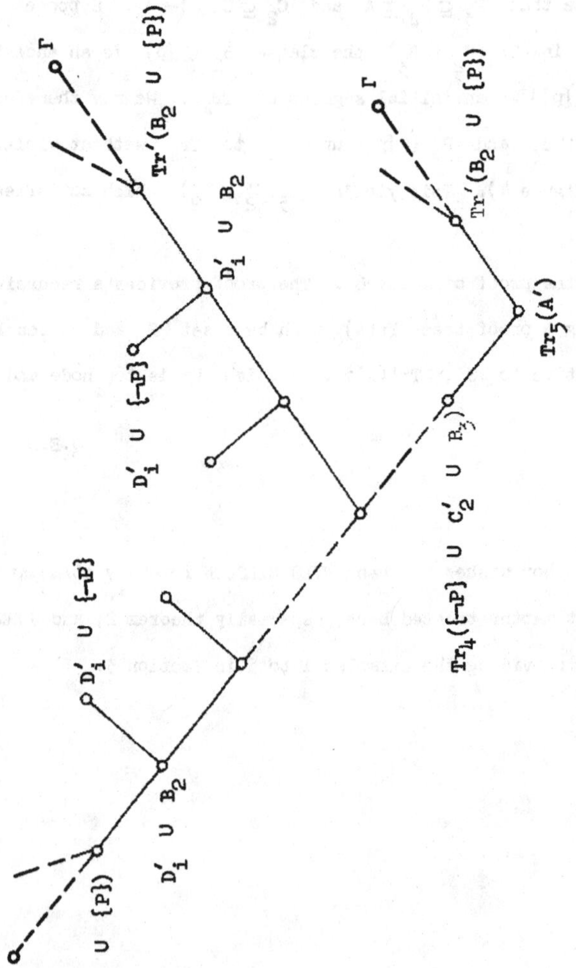

Proof of Lemma 6.

Figure 4

Footnotes

1. We will use the notation and terminology of references [2] and [4].

2. The disjunction connective is omitted from the clauses.

3. The conjunction of these two refinements has recently been proved complete (R. Kieburtz and D. Luckham, "Compatibility of Refinements of the Resolution Principle", forthcoming).

References

[1] Andrews, P. B., "Resolution with Merging", JACM, 15, No. 3, pp. 367-381, July 1968.

[2] Luckham, D., "The Resolution Principle in Theorem-Proving", Machine Intelligence 1, Collins and Michie (Eds), American Elsevier Inc., New York, 1967.

[3] Luckham, D., "Some Tree-Paring Strategies for Theorem-Proving", Machine Intelligence 3, D. Michie (Ed), Edinburgh University Press, pp. 95-112, 1968.

[4] Robinson, J. A., "A Machine-Oriented Logic Based on the Resolution Principle", JACM, 12, No. 1, pp. 23-41, January 1965.

[5] Robinson, J. A., "Automatic Deduction with Hyper-Resolution", International Journal of Computer Mathematics, Vol. 1, pp. 227-234, 1965.

[6] Slagle, J. R., "Automatic Theorem-Proving with Renamable and Semantic Resolution", JACM, 14, pp. 687-697, October 1967.

[7] Wos, L., Robinson, G., Carson, D., "Efficiency and Completeness of the Set of Support Strategy in Theorem Proving", JACM, 12, No. 4, pp. 536-541, October 1965.

[8] Wos, L., Robinson, G., Carson, D., Shalla, L., "The Concept of Demodulation in Theorem-Proving", JACM, 14, No. 4, pp. 698-704.

[9] Wos, L., and Robinson, G., "The Maximal Model Theorem", Abstract, Spring 1968 meeting of Association for Symbolic Logic, to appear in J. Symb. Logic.

DEFINITIONAL APPROACH TO AUTOMATIC DEMONSTRATION

If one speaks about the application of a computer to numerical calculations we understand exactly the idea he has in mind. But speaking about theorem proving by means of a computer seems to be not clear enough. This raises many disscusions on automatic demonstration - some times caused by misunderstanding the notion of a computer and /or the task which the computer has to perform by doing theorem proving. The aim of the presemted note is to give the main fields where the computer can be used as an instrumental aid in mathematical creativity. First we shall define the notion of a computer.

Let T be finite or infinite set and let π be a partial function $\pi : T \rightarrow T$. Sequence $t_0 , t_1 ,\ldots$ such that for all i , $t_i \in T$ and $t_{i+1} = \pi (t_i)$ will be called process. The process $t_0 ,\ldots, t_k$ is called finite if $t_k \notin D$, where D_π denotes the domain of the function π . Let us introduce binary relation $M \subset T x T$ defined as follows: $\langle t,t' \rangle M$ if and only if there exist finite process $t_0 ,\ldots, t_k$ such that $t_0 = t$ and $t_k = t'$. It is easly to show that relation M is a function. This function is called computer. T is refered to as a memory of a computer M and π is called the control of a computer M . It can be easly shown that each digital computer may be presented as a function M . We shall say that the computer M computes the function $f : X \rightarrow X$ if and only if for all $x \in X$, $f(x) = \delta \{M[\ (x)]\}$, where $: X \rightarrow T$ and $\delta : T \rightarrow X$ are called coding and decoding functions respectively. For the sake of simplicity we shall omit the coding and decoding functions and writte $y = M(x)$, which is to mean that we supply data x to the computer M and as a result of computation we obtain y .

In a similar way we may define the main tasks of computer application in automatic demonstration. Before we define the fields where the computer can be used in theorem proving let us first introduce some notations:

S - sentence in mathematical language

T - theorem

P_t - proof of the theorem T

$Cn\ T$ - the set of all consequences of T /by fixed set of axioms/

$Pr\ T$ - set of all premisses of T

Z - set of sentences

Ak - set of axioms.

By means of the above notations we can introduce the following definitions:

1 - Computation of truth value

$$M(S) = \begin{cases} 1 \text{ , if } S \text{ is a theorem,} \\ 0 \text{ , if } S \text{ is not a theorem.} \end{cases}$$

2 - Production of formal proof

$$M(S) = \begin{cases} P_S \text{ , if } S \text{ is a theorem,} \\ 0 \text{ , if } S \text{ is not a theorem.} \end{cases}$$

3 - Search for semantic proof

$$M(S,I) = \begin{cases} 1 \text{ , if } S \text{ is truth by the interpretation } I \text{ ,} \\ 0 \text{ , if } S \text{ is not truth by the interpretation } I \text{ .} \end{cases}$$

4 - Production of counterexample

$$M(S) = \begin{cases} 1 \text{ , if } S \text{ is a theorem} \\ \text{counterexample, if } S \text{ is not a theorem.} \end{cases}$$

5 - Production of concequences

$$M(Ak,T) = C \subset Cn(Ak,T) \text{ .}$$

6 - Production of premisses

$$M(T) = P \subset Pr(T) \text{ .}$$

7 - Investigation of independece

$$M(T,T') = \begin{cases} 0 \text{ , if } T \text{ and } T' \text{ are independent} \\ 1 \text{ , if } T \text{ and } T' \text{ are not independent.} \end{cases}$$

8 - Simplification of proof

$$M(T,P) = P'$$

9 - Verification of proof

$$M(T,Z) = \begin{cases} 0 \text{ , if } Z = P \\ 1 \text{ , if } Z \neq P \end{cases}$$

10 - Equivalence of axioms

$$M(Ak) = Ak' \text{ ,}$$

where Ak and Ak' are equivalent sets of axioms and Ak' is in some sence simpler then Ak .

It seems that these are the main tasks which can be soloved by means of a computer in automatic demonstartion. It would be interesting to discuss which of these tasks are most important in mathematical reasoning and which are most promissing in sucessfull computer application. If it turns out that there is a gap between these two fields the quastion arises how to bridge the gap.

This seems to be one of the most important problems in developing automatic demonstartion. According to my opinion the future of the application of computers in mathematical work lies not in batch processing but in conversational mode of using the computer in theorem proving. In other words it means that points 1 and 2 are less promissing then points 5 or 6 for example. There is a litlle hope that the computer can be usefull in production the whole proof for some theorem. We may rather expect some positive results by application of a machine to produce some partial results in the process of theorem proving-which may approach the main problem the whole proof of a theorem.

HEURISTIC INTEREST OF USING METATHEOREMS

Jacques PITRAT

Chargé de Recherches au C.N.R.S.

A - GOALS OF THE PROGRAM

I had two goals :

1. Write an autonomous program : it must prove interesting theorems and we do not give it what theorems it must prove. It only receives a method for computing the interest of a theorem. It must find theorems with as great as possible an interest.

   Hao Wang (1) tried to write such a program. It was based on a combinatorial generation of formulas, but he has some difficulties because there are many theorems.

2. Write a general program. It is not limited to one formal system. It receives as data the axioms and the rules of derivation of a formal system. Then it must study it.

   My program is general, because it could study several formal systems (there are is (2) the results for six formal systems). But it cannot study any formal system. And for some others, its performances are poor.

B - THE METATHEOREMS

The metatheorems are statements on a formal system. We must not confuse them with theorems, which are the statements of a formal system which we can deduce from the axioms by using the rules of derivation.

There are many kinds of metatheorems. Some indicate that all the elements of a set of formulas are theorems. For instance :

"All the formulas : $\supset \sim^{2n} p\, p$ are theorems in such formal system of the prepositional logic".

Others give new means for obtaining theorems. These means are not the rules given at the beginning, but are deduced from the rules and from the axioms.

For instance :

"If T is a theorem, we can obtain a new theorem by substituting to a term of T , this term preceded by two negations".

A new production is an example of such a metatheorem. A production is a rule of derivation :

"If A_1 is a theorem and if A_2 is a theorem... and if A_n is a theorem, then B is a theorem".

We write this also :

$$A_1 , A_2 , \ldots A_n \Rightarrow B$$

The A_i are the antecedents of the production.

The program can only prove metatheorems which are new productions. We will see why it is interesting to prove new productions, then how we can prove them.

We suppose that the formal systems given have the rule of substitution : we can substitute to a variable any term, but we must do the same substitution for all the occurrences of the variable. If the first formula is a theorem or a production, the formula obtained after the substitution is also a theorem or a production.

C - WHY THE PROGRAM USES NEW METATHEOREMS.

If we use metatheorems, we do not change the set of formulas which are theorems. But we can prove more easily the theorems which are interesting. There are two reasons :

1. It is easier to define the interest of a production than the interest of a theorem.

It is useful to give to each formula (theorem or production) an interest which has a great heuristic value. It is used many times by the program. For instance :

- if the interest of a result is lower than a threshold, this result is eliminated.

- we make the derivations using first the results with the highest interest.

This may be a very good method. When we eliminate results, we limit the number of trials, thus we avoid the combinative approach. This is based on experience : if the interest is well defined, the results with a low interest are rarely useful for getting the desired theorem. Also, the derivations using results with a high interest give very often new interesting results.

But this is true only if the interest is well defined. It is difficult to define the interest of a theorem. We can use the number and the variety of connectives and of variables. But if we want to have a definition of

The interest which is very good, we must use semantics : we must see what is the meaning of the theorem in some interpretation of the formal system. This is very difficult to do with a general program.

On the other hand, it is easy to give a general method for computing the interest of a production with one antecedent. Let $A \Rightarrow B$ be such a production. We have two strings A and B . We can compare them. Experience shows that more these strings are alike, more the interest of the production is high. Also, they must not be too complicated. It is easy to define the resemblance of two strings and to have a good definition of the interest of a production with one antecedent.

Suppose we do not find new productions, and we have a formal system with only one production with two antecedents (like modus ponens in many formal systems of the propositional logic) . We must choose two theorems : we must unify one with the first antecedent, and the other with the second antecedent of the production.

But if we have proved new productions with one antecedent, we must choose ont theorem and one production.

In the first case, we have to choose two results the interest of which is poorly defined. In the second case, the interest of the theorem is not well defined, but the interest of the production is very well evaluated. We have better performances because we have to choose only one result with an interest difficult to define.

This is the first reason.

2. <u>When we prove a new production, we memorize that a sequence of derivations is useful.</u>

Suppose we have a formal system which has the following axioms (written in polish prefixed notation)

$$A1 \quad \supset\supset p \supset q\, r \supset\supset p\, q \supset p\, r$$

$$A2 \quad \supset p \supset q\, p$$

and the production : $p \supset p\, q \Rightarrow q$

Suppose we have the theorem $T0$: $\supset P \supset Q R$, P, Q, R are terms, which we do not specify now.

When we describe a derivation, we indicate :

- the name of the theorem which we unify with the first antecedent,
- the name of the theorem which we unify with the second antecedent,
- the theorem resulting from the derivation,
- the name of this new theorem

$T0$	$A1$	$\supset\supset P Q \supset P R$	$T1$
$T1$	$A2$	$\supset s \supset\supset P Q \supset P R$	$T2$

is a variable which does not occur is P, Q, R.

$T2$	$A1$	$\supset\supset s \supset P Q \supset s \supset P R$	$T3$
$A2$	$T3$	$\supset Q \supset P R$	$T4$

We see that if $\supset P \supset QR$ is a theorem, then $\supset Q \supset PR$ is a theorem. We proved the production :

$$\supset p \supset q r \Rightarrow \supset q \supset p r$$

This production is very interesting :

With $\supset \supset p q \supset \supset r p \supset r q$ we obtain $\supset \supset p q \supset \supset q r \supset p r$

$\supset p p$ $\qquad$ $\supset p \supset \supset p q q$

$\supset \sim p \supset p q$ $\qquad$ $\supset p \supset \sim p q$

$\supset \supset p \supset q r \supset \supset p q \supset p r$ $\qquad$ $\supset \supset p q \supset \supset p \supset p r \supset p r$

If we try to prove $\supset \supset p q \supset \supset q r \supset p r$ from $\supset \supset p q \supset \supset r p \supset r q$ without using the new production, but doing the same demonstration, we have three intermediary results corresponding to $T1$, $T2$ and $T3$.

as P is $\supset p q$

Q is $\supset r p$

R is $\supset r q$

We see that these results are :

$$\supset \supset \supset p q \supset r p \supset \supset p q \supset r q$$

$$\supset s \supset \supset \supset p q \supset r p \supset \supset p q \supset r q$$

$$\supset \supset s \supset \supset p q \supset r p \supset s \supset \supset p q \supset r q$$

These three results have a low interest. We have two possibilities :

- We store few results : the threshold is high. We will never obtain the result wanted : $\supset \supset p q \supset \supset q r \supset p r$, because we will eliminate these intermediary results.

- we store many results : the threshold is low. We will keep these three intermediary results if we prove them. But we will keep also many other results with a low interest. Besides, for all these intermediary results, we will try them in many derivations, although only one derivation is useful ; but we do not know which beforehand. The others will give results with a low interest which are not useful as intermediary results in order to find some interesting result. But it is not possible to see that only by looking at them ! We will spend a lot of time with these results and with the theorems obtained from these results.

We see two consequences :

- the storage will be too small
- the computing time will be very long.

If we look at the tree developed, we will have :

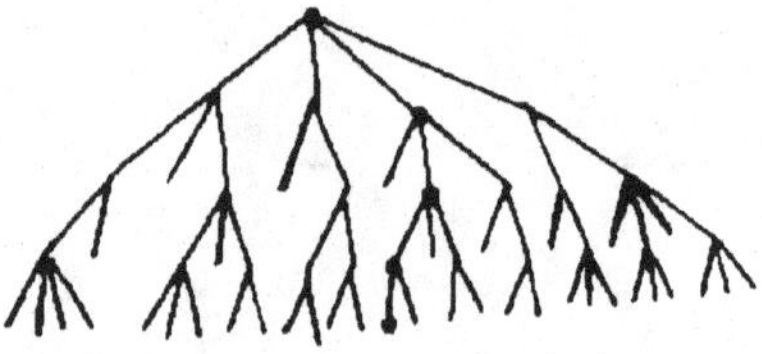

When we used the production with one antecedent, we had :
and we do not even need to write the intermediary results.

One could say : it is not necessary to prove : $\supset p \supset q r \Rightarrow \supset q \supset p r$
but we could wait to prove : $\supset \supset p \supset q r \supset q \supset p r$, Unfortunately, in many formal system, the proof of this theorem is easy to find only when we have found the production first.

The second reason for proving new productions is : we memorize that some sequence of derivations is useful. We do not make the same derivation many times and it is not necessary to keep the intermediary results with a low interest. We can have the threshold (under which we eliminate the results) very high.

- HOW TO OBTAIN THE METATHEOREMS

When we write a program applicable to one formal system, the author of the program can prove some metatheorems and use them. For instance, the Logic Theorist (3) uses :

$$\supset p\, q\,,\ \supset q\, r \Rightarrow\ \supset p\, r$$

which is not included in the rules given at the beginning in the Principia. It is very easy to prove in this formal system, and the authors are right when they use it. It is one of the most important rules of the program.

But in some cases we must use metatheorems, the proof of which is difficult to find. If they must be proved by a man, what is the use of having a theorem proving program ? Besides, the metatheorems are not the same for any one formal system. A metatheorem may exist in one formal system and not in another. Therefore we must find another solution when we write a general program.

The program must prove the metatheorems which it will use thereafter. As my program is autonomous, it is not difficult to implement. The same heuristic methods which are useful to induce it to find interesting metatheorems. The program will try to finc interesting theorems and interesting metatheorems and we do not indicate to it which results it may find.

The only difference between proving theorems and metatheorems is the following. When it proves theorems, it uses metatheorems which are means to obtain new theorems. When it proves metatheorems, it uses metametatheorems, which are means to obtain new metatheorems.

Such a metametatheorem is :

"If $a \Rightarrow b$ is a production and if $b \Rightarrow c$ is a production, then $a \Rightarrow c$ is a production".

Knowing two productions, we can get a new production. We can note that this metametatheorem is general and useful for any formal system. It formulates the transitivity of " $\Rightarrow$ ".

When we know the productions :

$$\supset p \supset q r \Rightarrow \supset \supset p q \supset p r$$

$$\supset \supset p q r \Rightarrow \supset q r$$

after applying the metametatheorem, we obtain the new production :

$$\supset p \supset q r \Rightarrow \supset q \supset p r$$

It would be very interesting if all the metametatheorems were general. It would be sufficient to give them to the program with heuristics indicating when we can use them.

Unfortunately some metametatheorems are not general, for instance :

"If $\supset p q$ is a theorem, then $p \Rightarrow q$ is a production"

We see a connective of the formal system : $\supset$ and it cannot exist in a formal system which has not this connective. It is not obvious that it is a metametatheorem, even for the formal system which have this connective.

With this metametatheorem, we have a new production obtained from every theorem beginning with "$\supset$" . For instance, from : $\supset \supset p \supset q r \supset \supset p q \supset p r$, we can deduce : $\supset p \supset q r \Rightarrow \supset \supset p q \supset p r$

Another metametatheorem useful in some formal systems is :

"If $p \Rightarrow \supset q r$ is a production, then $p, q \Rightarrow r$ is a production".

From a production with one antecedent, we can deduce a production with two antecedents. But we must prove this metametatheorem before we can use it. There is also a connective of the language : "$\supset$" .

If the program has to prove metametatheorems, it mus know metametametatheorems which are means to obtain new metametatheorems. We could think that there is no reason to stop in this hierarchy of metalanguages. Fortunately, all the necessary metametametatheorems are general and useful for any formal system. There is no connective of the language, only connectives of the metalanguage such as : " $\Rightarrow$ " or " , ".

Here are some metametametatheorems :

"If $a, b \Rightarrow c$ is a production, we have the metametatheorem :

"If b is a theorem, then $a \Rightarrow c$ is a production"

If the production is : $p, \supset pq \Rightarrow q$ we obtain the metametatheorem :

"If $\supset p q$ is a theorem, then $p \Rightarrow q$ is a production".

But if the production is $p, |p| q r \Rightarrow r$ ($|$ is the Sheffer's stroke), we obtain the metametatheorem :

"If $|p| q r$ is a theorem, then $p \Rightarrow r$ is a production".

Another metametametatheorem is :

"If $a, b \Rightarrow c$ is a production, we have the metametatheorem :

"If $p \Rightarrow b$ is a production, then $p, a \Rightarrow c$ is a production".

is a variable which does not occur in a, b, c ".

When the production is : $p, \supset p q \Rightarrow q$, we obtain :

"If $p \Rightarrow \supset q r$ is a production, then $p, q \Rightarrow r$ is a production".

When the production is : $p, |p| q r \Rightarrow r$, we obtain :

"If $p \Rightarrow |q| r s$ is a production, then $p, q \Rightarrow s$ is a production".

I give in (2) all the metametatheorems and metametametatheorems used by the program. All are perfectly general. I indicate also how the program chooses to apply such a metametatheorem or such a metametametatheorem.

E - CONCLUSION

A theorem proving program must use metatheorems, if we want an efficient program. We have seen why it is interesting to use metatheorems. In fact, the mathematician uses them also.

A general program cannot receive the metatheorems at the beginning : they are applicable only to one formal system. It must prove them, and for this it uses metametatheorems. Some of these are general, and we give them to the program. But some are specific. We cannot give them to the program, we give it instead some possibilities to find them. These possibilities are metametametatheorems. Fortunatèly, all those which are necessary are general. It is not necessary to consider higher metalinguistic levels.

Some results obtained by the program which uses these methods are given in (2) for six formal systems. In the formal system with the axioms :

$$\supset \supset p \supset q r \supset \supset p q \supset p r$$

$$\supset p \supset q p$$

$$\supset \supset \sim p \sim q \supset q p$$

and the production : $p, \supset pq \Rightarrow q$, it proved for instance $\supset \supset p p p$.

In the formal system with the axiom :

$$\supset \supset \supset p q r \supset \supset r p \supset s p$$

and the production : $p, \supset p q \Rightarrow q$, it proved $\supset \supset \supset p q p p$.

(For these two formal systems, it proved many other theorems).

It is interesting to write a program doing metatheory in other areas than theorem proving. It is useful also for a general problem solver. Such a program must study the formulation of the particular problem which it must solve.

In (4), I try to apply these ideas to a general game playing program.

BIBLIOGRAPHY

1. HAO WANG Toward mechanical mathematics.
I B M Journal of Research and Development
Vol. 4 N° 1 Jan. 60 p. 2.

2. PITRAT Réalisation de programmes de démonstration de théorèmes utilisant des méthodes heuristiques.
Thèse 4 mai 1966.

3. NEWELL, SHAW, SIMON Empirical exploration with the logic theory machine ; a case study in heuristics.
Computers and thought
Editors Feigenbaum and Feldman. Mc Graw-Hill.
1963 p. 109.

4. PITRAT Realization of a general game playing program.
I F I P Congress 1968 Booklet H p. 120.

A PROOF PROCEDURE WITH MATRIX REDUCTION

Dag Prawitz

Introduction

I shall develop somewhat further a proof procedure for predicate logic that I have suggested earlier, and shall try to make some comparisons with other proof methods, in particular the so-called resolution procedure. The procedure that I shall describe is a modification of one that I suggested in 1967 (published in Prawitz 1969), which in turn is a development of a procedure proposed in Prawitz 1960.

The general form of the procedure

I shall consider complete methods for shoving that a set Γ of formulas in predicate logic is inconsistent. The formulas are supposed to be formulated in the language of first order predicate logic with function symbols. As is well-known (using Skolem-functions), there is no loss in generality to assume that the formulas of Γ are of the form

$$\forall x_1 \ \forall x_2 \ \dots \ \forall x_k \ (A_1 \vee A_2 \vee \dots \vee A_m)$$

where every A_i is an atomic formula or the negation of such a formula.

All complete proof methods that have been suggested (possibly with the exception of the one proposed by Professor Kanger at this conference) relies upon Herbrand's theorem, which says in this context that Γ is inconsistent if and only if there is a finite, inconsistent set of instances of the formulas of Γ obtained by substituting constant terms for the quantified variables. By a constant term is then meant a term build up of individual constants and function symbols occurring in some formula of Γ (or if there is no such constant, we consider the terms build up function symbols occurring in some formula of Γ and some arbitrarily chosen individual constant).

Herbrand's theorem gives rise to the following procedure for showing Γ inconsistent : Enumerate the instances of the formulas of Γ in some arbitrary order and form increasingly larger sets of these instances until an inconsistent one is found. The first mechanical proof procedures that wera developed (those by Gilmore, Prawitz and Wang around 1960) relied exclusively upon this idea.

It was soon realized, however, that an efficient proof procedure had to use some less crude method for generating substitution instances of the given formulas. A strategy for finding the substitutions that might lead to an inconsistent set of instances was suggested by Kanger and myself in 1959-60 and is generally used in one form or the other in the proof procedures considered today.

The basic strategy

I shall give a rough description of this strategy (a more complete description may be found in Prawitz 1960). To find a finite, inconsistent set of instances of the formulas of Γ is the same as to find (i) a *matrix* M of the form described below and (ii) a *refutation substitution* S to the matrix M as defined below. The matrix M is to be of the form

$$\begin{array}{l} A_{1,1}\ ,\ A_{1,2}\ ,\ldots,\ A_{1,m_1} \\ A_{2,1}\ ,\ A_{2,2}\ ,\ldots,\ A_{2,m_2} \\ \cdot \\ \cdot \\ \cdot \\ A_{n,1}\ ,\ A_{n,2}\ ,\ldots,\ A_{n,m_n} \end{array}$$

where for each line i , the disjunction $(A_{i,1} \vee A_{i,2} \vee \ldots \vee A_{i,n_i})$ is an alphabetic variant of the sentential part of a formula of Γ . Variables in different lines are supposed to be distinct. By a *path* in M , we mean a sequence of formulas obtained by taking one formula from each line of M ; i.e. a path is a sequence of the form $A_{1,j_1}\ ,\ A_{2,j_2}\ ,\ldots,\ A_{n,j_n}$ (where $j_i \leq m_i$ for any $i \leq n$). A substitution S of terms for the variables occurring in M is said to be a *refutation substitution* to M if it makes each path of M to contain a contradictory pair $Pt_1\ t_2 \ldots t_n$ and $\sim Pt_1\ t_2 \ldots t_n$. It is easily checked whether there exists a refutation substitution to a matrix M (without generating substitutions in some arbitrary order). One may, e.g., impose certain conditions on S, which, if fulfilled, make S a refutation substitution. S is said to fulfill $t = u$, if the substitution S makes the terms t and u identical. One may build up a set of conditions of the form $t = u$ by going through the different paths of M, for each path picking out a pair $Pt_1\ t_2 \ldots t_n$ and $\sim Pu_1\ u_2 \ldots u_n$, and forming the conditions $t_1 = u_1$, $t_2 = u_2, \ldots, t_n = u_n$. A substitution that fulfills all conditions in such a set is obviously a refutation substitution. If, when building up the set of conditions, we come to a point where no substitution can fulfill all the conditions (e.g., if the set contains two conditions $f(x) = y$ and $y = g(z)$, then for some path, we have to pick out another pair $Pt_1 t_2 \ldots t_n$ and $\sim Pu_1\ u_2 \ldots u_n$ and instead include the corresponding conditions obtained from this pair in the set of conditions. To find an inconsistent set of instances of the formulas of Γ, we may thus form successively larger matrices testing for each one whether there exists a refutation substitution until such a substitution is found.

Matrix reductions

The method to find a refutation substitution outlined above, requires that one runs through all paths of the matrix. However, a pair of formulas $A_{i,j}$ and $A_{p,q}$ $(i \neq p)$ picked out from one path of the matrix belong also to several other paths of the matrix, and having imposed a condition on S that makes $A_{i,j}$ and $A_{p,q}$ contradictory we should not need to consider all these other paths. To this end, I shall describe an operation that I shall call matrix reduction. Let $A_{i,j}$ and $A_{p,q}$ be two formulas from different lines (i.e., $i \neq p$) in the matrix M that can be made contradictory by some substitution. By a *matrix reduction of* M *with respect to* $A_{i,j}$ and $A_{p,q}$, I then mean the operation of first performing a most general substitution in M that makes $A_{i,j}$ and $A_{p,q}$ contradictory and then replacing the obtained matrix M' as follows :

(i) If none of $A_{i,j}$ and $A_{p,q}$ is alone on its line (i.e., n_i, $n_q > 1$), then M' is replaced by the two matrices M_1 and M_2, where M_1 is obtained from M' by leaving out first all formulas on line i except $A_{i,j}$ and then the formula $A_{p,q}$, and where M_2 is obtained by leaving out $A_{i,j}$ and all formulas of the line p except $A_{p,q}$.

(ii) If one but not both of $A_{i,j}$ and $A_{p,q}$ is alone on its line, then M' is replaced by M_0 obtained by leaving out that formula of $A_{i,j}$ and $A_{p,q}$ which is not alone on its line.

(iii) If both of $A_{i,j}$ and $A_{p,q}$ are alone on their lines, no replacement is made.

In case (i), we say that we have a *matrix split;* in case (ii), we have a *simple reduction;* and in case (iii), we say that M' is *inconsistent*.

Clearly, a substitution S that makes $A_{i,j}$ and $A_{p,q}$ contradictory is a refutation substitution to M if case (iii) applies, and if case (i) or (ii) applies, then S is a refutation substitution to M if and only if it is a refutation substitution to both M_1 and M_2 or, respectively, it is a refutation substitution to M_0.

the matric reduction can be somewhat refined when case (i) applies by making the variables that occur in other lines in M_2 than i and p distinct from the variables in M_1.

The procedure of matrix reduction

By a matrix split or a simple reduction we reduce the problem of finding a refutation substitution to M to the same problem for two simpler matrices or one simpler matrix, respectively. For instance, if M is a matrix with 4 lines and 4 formulas on each line, and hence has $4^4 = 256$ paths, then a matrix split will give two matrices, each one with only $3 \cdot 4^2 = 48$ paths; i.e., instead of having to consider 256 paths, we have only to consider 96 paths after one matrix reduction. To find a refutation substitution, we have only to carry out a series of matrix reductions until we get only inconsistent matrices. Such a series of matrix reductions can be carried out in different ways. One way is as follows.

Let M be the matrix to which we want to find a refutation substitution. Let C_0 be $\{M\}$ and let S_0 be $\{C_0\}$. If C is any set of matrices, then by a reduction of C we understand a set C' obtained by replacing one of the matrices in C by the result of a matrix reduction, carrying out the corresponding substitution also on the variables occurring in the other matrices of C . We may then construct a series of sets S_0 , S_1 , S_2 ,... by letting S_{i+1} be $\{C' : C'$ is a reduction of some $C \quad S_i\}$. We continue this series until we get an S_p which either contains a set C of inconsistent matrices or contains only sets C which cannot be further reduced. In the first case, we have found a refutation substitution to M , and in the second case, there is no refutation substitution to M . In the second case we have to extend the matrix M and construct a new series S_0' , S_1' , S_2' ,... starting with the extended matrix M'. Obviously, we can then use the series S_0 , S_1 , S_2 ,...,S_p already obtained in order to facilitate the construction of S_0' , S_1' , S_2' ,... In this respect, the procedure that has just been described differs from the one given in Prawitz 1969. There are some unnecessary repetitions in the procedure as described above; e.g., there may occur some series of operations that differ only in the order in which the operations are carried out, and such repetitions should of course be eliminated.

To show that Γ is inconsistent, we may start with a matrix M , where the sentential part of each formula of Γ occurs once as a line of M . Each time that we have to extend the matrix, we may then add one line for each formula of Γ , i.e., we take an alphabetic variant of the sentential part of each formula of Γ as a new line. But other ways of extending the matrix is of course also possible.

Some comparisons with the resolution procedure

The proof procedure developed here seems to be similar to one recently suggested by Loveland 1968. However, his procedure is formulated rather differently and I do not know the exact relationship between the two procedure.

I shall conclude by indicating how the procedure suggested here differs from the resolution procedure developed by Robinson 1965.

They differ in three main respects. Firstly, instead of the operation of matrix reduction, the resolution procedure uses the resolution operation. This operation operates on two formulas $(A_1 \, v \, A_2 \, v \, \dots \, v \, A_n)$ and $(B_1 \, v \, B_2 \, v \, \dots \, v \, B_n)$ and consists of first, finding a substitution that makes a pair A_i and B_j contradictory and then, after having performed this substitution, forming the disjunction of all the A's and B's leaving out A_i and B_j (and any A_k or B_k that after we substitution becomes identical to A_i or B_j, respectively). Thus, while matrix reduction is based on the equivalence between $(A \, v \, B)$ & $(\sim A \, v \, C)$ and $(A \,\&\, C) \, v \, (B \,\&\, \sim A)$, resolution is based on the fact that $(A \, v \, B)$ & $(\sim A \, v \, C)$ implies $B \, v \, C$. Since $B \, v \, C$ is a weaker proposition than $(A \,\&\, C) \, v \, (B \,\&\, \sim A)$, it is in general more difficult to find an inconsistency between $B \, v \, C$ and some other propositions than between $(A \,\&\, C) \, v \, (B \,\&\, \sim A)$ and some other propositions. This may indicate that one looses some information by carrying out a resolution, which may have to be recovered by some other resolutions, but which is retained in a matrix reduction.

Secondly, there is a difference between the two procedures which may be indicated by saying that in the matrix reduction procedure as described above, one considers just one way of extending the matrix when no refutation substitution is found, while the resolution procedure considers all ways in which the matrix can be extended, so to say. The matrix reduction procedure can easily be varied in this respect but it is doubtful whether the fact that the resolution procedure may found a sherter matrix to which there exists a refutation substitution, balances the considerable extra work involved in considering all ways of extending a matrix.

Thirdly, the resolution procedure allows resolution also on two formulas that have been obtained by preceding resolutions. The matrix reduction procedure could be modified so that it becomes similar to the resolution procedure in this respect by allowing combinations of matrices belonging to different C's in a set S_i. Whether this would be an advantage is difficult to know but there is some evidence for that.

Finally, the resolution procedure has been amplified by the addition of several refined strategies. They usually do not essentially depend on the resolution principle, however, but could be added also to the matrix reduction procedure.

Although there can be no doubt about the greater efficiency of, e.g., the basic strategy for finding a substitition which gives an inconsistent set of instances as compared to the crude method of just enumerating all instances, it seems difficult to evaluate the differences between two procedures such as the resolution method and the matrix reduction method. Some measure for evaluating different procedures is therefore strongly needed.

BIBLIOGRAPHICAL REFERENCES

Loveland (1968), D.W., Mechanical theorem proving by model elimination, *Jour. ACM 15,* pp. 236-251.

Prawitz (1960), Dag, An improved proof procedure, *Theoria 26,* pp. 102-139.

Prawitz (1967), Advances and problems in mechanical proof procedures, to appear.

Robinson (1965), J.A., A machine oriented logic based on the resolution principle, *Jour. ACM 12,* pp. 23-44.

AXIOM SYSTEMS IN AUTOMATIC THEOREM PROVING*

George Robinson • Lawrence Wos

In judging the suitability of axiom systems for automatic theorem proving in a particular domain, two considerations are of prime importance: *logical strength* and *proof-search efficiency*. The axiom system should be just strong enough to yield all the theorems of the theory and only the theorems of the theory. We shall employ the concept of *adequacy* (defined in the following section) to study this property. Given two axiom systems each of appropriate strength for the theory in question, one would prefer to use the system that would, when used by the theorem-proving programs in question, discover proofs in the shortest time (or using the smallest amount of memory or other computer resources). Examples showing how the choice of axiom system can have important effects on proof-search will be given in the section on efficiency.

*Work performed under the auspices of the U. S. Atomic Energy Commission.

We shall be dealing with a language equipped with denumerably many individual variables, individual constants, n-adic function constants, and n-adic predicate constants. An individual constant or individual variable is a *term*, as is an n-adic function constant followed by n terms. A *ground term* is one involving no variables. An *atomic formula* (or *atom*) is an n-adic predicate letter followed by n-terms. A *literal* is an atomic formula or the negation thereof. A *clause* is the disjunction of finitely many literals; it is a *ground clause* if no variables occur. It is often profitable to identify clauses, particularly ground clauses, with the set whose members are the literals occurring in the clause.

If some or all of the variables of a clause are systematically replaced by terms, the resulting clause is an *instance* of the original clause.

A set of ground literals (literals involving no variables) *satisfies* a ground clause if it has a non-empty intersection with the clause; it *satisfies* any clause (with respect to some herbrand universe, H_S) if it satisfies all ground instances (over H_S) of that clause. If a set of literals contains the negation of each literal of a ground clause, it *condemns* the ground clause; if it condemns some ground instance of any clause, it *condemns* the clause. The clause □ (read "the empty clause" or just "nil") having no literals can be satisfied by no set of literals and is (vacuously) condemned by every such set.

The *herbrand universe*, H_S, of a set S of clauses is the set of all ground terms that can be formed from the functions and individual constants occurring in S, with the constant a being supplied in the case that S contains no indidual constants. An *herbrand atom* for S is an atomic formula composed of a predicate letter occurring in S and the appropriate number of terms from H_S. An *interpretation* I of S over H_S is a set of (ground) literals such that for each herbrand atom L for S, exactly one of L or $\sim$L is in I. A *model* of S (over H_S) is an interpretation of S over H_S that satisfies S.

If S and T are sets of clauses, the S *implies* T ($S \models T$) if no model of S condemns T. Then each clause C in T is a *logical consequence* of S (is *implied by* S, $S \models C$).

This paper concerns itself principally with sets of clauses. There appears to be nc reason, in principle, why the discussion should not go over to more general sentences, such as those obtained by using logical operators for conjunction, conditional, biconditional, and more general application of negation and quantification. (In effect, with clauses one considers only universal quantification over each clause.) In the discussion of eliminability, the biconditional $\equiv$ is in fact used, since the statement of this property appears to become inordinately complex in terms of clauses alone. We assume some conventional rules of function, inference, and semantics for these additional connectives.

For clauses, three rules of inference are of prime importance in automatic theorem proving: *factoring, paramodulation,* and *resolution.*

Definition (Paramodulation): Let A and B be clauses such that a literal Rst (or Rts) occurs in A and a term u occurs in (a particular position in) B. Further assume that s and u have a most general common instance $s' = s\sigma = u\tau$ where σ and τ are the most general substitutions such that $s\sigma = u\tau$. Where $\hat{B}$ is obtained by replacing by $t\sigma$ the occurrence of $u\tau$ in the position in $B\tau$ corresponding to the particular position of the occurrence of u in B, infer the clause $C = \hat{B} \cup (A - \{Rst\})\sigma$ (or $C = \hat{B} \cup (A - \{Rts\})\sigma$). C is called a *paramodulant* of A and B and is said to be *inferred by paramodulation from* A *on* Rst (or Rts) *into* B *on (the occurrence in the particular position in* B *of)* u. The literal Rst (or Rts) is called the *literal of paramodulation.* [4]

Definition: For any literal l, /l/ is that atom such that either l = /l/ or l = ∿/l/.

Definition (Resolution): If A and B are clauses with literals k and l respectively, such that k and l are opposite in sign (i.e., exactly one of them is an atom) but /k/ and /l/ have most general common instance m, and if σ and τ are the most general substitutions with $m = /k/\sigma = /l/\tau$, then infer from A and B the clause $C = (A - \{k\})\sigma \cup (B - \{l\})\tau$. C is called a *resolvent* of A and B and is inferred by *resolution.*

Definition (Factoring): If A is a clause with literals k and l such that k and l have a most general common instance m, and if σ is the most general substitution with $k\sigma = l\sigma = m$, then infer the clause $A' = (A - \{k\})\sigma$ from A. A' is called an *immediate factor* of A. The *factors* of A are given by: A is a factor of A, and an immediate factor of a factor of A is a factor of A.

Given a set S of sentences and implicitly understood rules of formation and semantics, we shall be interested in the set W(S) of all well-formed sentences over the vocabulary of S (with a denumerable supply of individual variables added if necessary) and the set $V(S) = \{A | A \in W(S), S \models A\}$. A theory T will be thought of as being defined by the set W(T) of well-formed sentences of the theory and the set V(T) of valid formulas of the theory even when we have no particular set of axioms in mind for T.

Adequacy of an Axiom System

A set E of sentences is a *non-creative** extension of a set E' if $V(E) \cap W(E') = V(E')$; it is an *eliminable* extension if for every $C \in W(E)$ there is a $C' \in W(E')$ such that $E \models C \equiv C'$.

In Figure 1 two (redundant) sets of axioms are given.** A1-8 were obtained by writing in clause form a set of axioms given by Abraham Robinson*** for group theory in terms of a single binary relation

*The concepts of non-creativity and eliminability as used here are closely related to the two criteria for definitions given in [7], page 154; hence the choice of terminology.

**{A1,...,A8}, {A6,...,A10}, and {A6,A7,A8,A9,A11,A12} are equivalent sets. {B1,B2,B5,B6,B7,B8,B9,B11,B13}, {B6,B7,B9,B11,B13,B14}, and {B6,B7,B11,B12,B13,B15,B16} are equivalent sets and each of them implies the three sets of A-axioms. Since A3 appears to be a more natural way of stating associativity one might ask whether A10 might not be replaced by A3 in the set A6-10. This question is answered negatively by the following counter-example:

Consider a domain of three elements {0,1,2}. Let f map to the usual cyclic group operation on this domain, let g map to the corresponding group inverse operation, and let e map to 0, but let R map to the equivalence relation K such that $0 \neq 1 = 2$; namely $K = \{(0,0), (1,1), (1,2), (2,1), (2,2)\}$. A3, A6, A7, A8, and A9 are all obviously satisfied, but A5 is falsified by the choice of $x = y = z = 1$ and $u = 2$, since $(1,1) \in K$, $(1,2) \in K$, and $(f(11),f(12)) = (2,0) \notin K$.

***Reference [3], p. 26.

A1	$\bar{R}xy\ Ryx$	(sym)	B1	$\bar{R}xy\ Ryx$
A2	$\bar{R}xy\ \bar{R}yz\ Rxz$	(trans)	B2	$\bar{R}xy\ \bar{R}yz\ Rxz$
A3	$Rf(f(xy)z)f(xf(yz))$	(assoc.)		
A4	$\bar{R}xy\ Rg(x)g(y)$	(g-subst.)		
A5	$\bar{R}xz\ \bar{R}yu\ Rf(xy)f(zu)$	(f-subst.)	B5	$\bar{R}xt\ \bar{R}yu\ \bar{R}zw\ \bar{P}xyz\ Ptuw$
A6	$Rf(ex)x$	(l.ident.)	B6	$Pexx$
A7	$Rf(g(x)x)e$	(l.inv.)	B7	$Pg(x)xe$
A8	Rxx	(reflex.)	B8	Rxx
A9	$\bar{R}xz\ \bar{R}yz\ Rxy$		B9	$\bar{P}xyz\ \bar{P}xyu\ Rzu$ (uniq.of prod.)
A10	$\bar{R}f(xy)u\ \bar{R}f(yz)t\ Rf(uz)f(xt)$			
A11	$\bar{R}f(xy)u\ \bar{R}f(yz)t\ \bar{R}f(uz)w\ Rf(xt)w$	(assoc.)	B11	$\bar{P}xyu\ \bar{P}yzt\ \bar{P}uzw\ Pxtw$
A12	$\bar{R}f(xy)u\ \bar{R}f(yz)t\ \bar{R}f(xt)w\ Rf(uz)w$	(assoc.)	B12	$\bar{P}xyu\ \bar{P}yzt\ \bar{P}xtw\ Puzw$
			B13	$Pxyf(xy)$ (closure)
			B14	$\bar{R}zu\ \bar{P}xyz\ Pxyu$ (P_3-subst)
			B15	$\bar{R}xy\ Pexy$
			B16	$\bar{P}exy\ Rxy$
			B17	$\bar{P}xyz\ \bar{P}xyu\ Pezu$
			B18	$\bar{P}ezu\ \bar{P}xyz\ Pxyu$
			B19	$\bar{R}f(xy)z\ Pxyz$
			B20	$\bar{P}xyz\ Rf(xy)z$
A21	$Rf(xe)x$	(r.ident.)	B21	$Pxex$
A22	$Rf(xg(x))e$	(r.inv.)	B22	$Pxg(x)e$

Figure 1.

R and two functions f and g. B1-2, B5-9, B11, and B13 were obtained from another set of group theory axioms in the same book**** by replacement (in a set of sentences not originally involving function symbols) of existentially-quantified variables by Skolem function symbols.

Consider the set $S4 = S1 \cup \{B15, B16\}$, where $S1 = \{B6, B7, B11, B12, B13\}$. Thus defined, S4 is obviously a non-creative, eliminable extension of S1.

If a theory T is a non-creative, eliminable extension of a set S of axioms, then the set S is of appropriate logical strength for a study of the theory T by means of automatic theorem-proving. That is, since every sentence C in W(T) can be mapped into a sentence $C\mu$ in W(S) in such a way that $C\mu$ is in V(S) iff C is in V(T), we need only apply a proof procedure to V(S). But confining the choice of axiom sets to those which have T as a non-creative, eliminable extension is unnecessarily restrictive. It forbids, for example, using a system such as S1 to study one such as A1-A8, when the former is much more efficient in proof search with some types of inference apparatus than the latter. This problem is not avoided by allowing the use of a set S which is itself a non-creative, eliminable extension of the theory. It appears appropriate for automatic theorem proving to make only a requirement concerning the existence of an appropriate (and effective) mapping μ. We shall do this by defining a set of sentences S to be *adequate* for a theory T iff there exists a uniform means of transforming the atomic formulae of W(T) into atomic formulae of W(S) such that the mapping μ induced on all sentences of W(T) maps the

****op. cit., p. 229.

sentences C in W(T) into sentences $C\mu$ of W(S) such that

$C\mu \in V(S)$ iff $C \in V(T)$.*****

Whenever T is a non-creative, (effectively) eliminable extension of S, S is adequate for T. By *effectively eliminable* we mean that there is an effective means of determining the C', given the C. In that case, let $C\mu = C'$ for $C \in W(T) - W(S)$ and let $C\mu = C$ for $C \in W(S)$. Then for $C \in W(S)$, $C\mu = C \in V(S)$ iff $C \in V(T)$ since $V(T) \cap W(S) = V(S)$. For $C \in V(T) - V(S)$, we have $C\mu \in V(T)$(since $T \models C \equiv C\mu$) and hence $C\mu \in V(S)$. For $C \in W(T) - V(T)$, we have $C\mu \notin V(S)$, since if $C\mu \in V(S)$ we would have $C\mu \in V(T)$ and hence $C \in V(T)$ (since $T \models C \equiv C\mu$) contrary to hypothesis.

On the other hand, S can be adequate for T but fail to be a non-creative extension of S as illustrated by the trivial example $T = \{P\}$, $S = \{\overline{P}\}$ (μ is the mapping from P to $\overline{P}$, and from $\overline{P}$ to P).

If we let S2 = {B9,B14} and consider S3 = S1 $\cup$ S2 we find that S1 is adequate for S3.****** To see that this is the case, let τ map each

*****It may be possible to generalize the concept of adequacy by partially relaxing the restriction that μ be induced by a transformation of atomic formulae. The restriction is quite natural for automatic theorem proving, since it requires, in effect, that $C\mu$ be a clause if C is a clause. The restriction apparently cannot be completely removed to allow arbitrary effective μ without admitting, for example, certain pathological mappings by which any set of tautologies in a sufficiently rich vocabulary could be shown adequate for any finitely-axiomatizable theory.

******Henceforth we shall restrict the discussion to consider only clauses as sentences, i.e., W(S) will always be a set of clauses.

atomic formula of the form $R\alpha\beta$ into $Pe\alpha\beta$ while leaving one of the form $P\gamma\alpha\beta$ unchanged. Then let θ be the mapping on clauses induced by τ. First we show that θ maps V(S3) into V(S1). Suppose by way of contradiction that for some C, S3 $\models$ C but S1 $\not\models$ Cθ. Then there is a model M of S1 over H_{S1} that condemns some ground instance Dθ of Cθ, where D is a ground instance of C. Let $M^* = M \cup \{R\alpha\beta | Pe\alpha\beta \in M\} \cup \{\overline{R}\alpha\beta | Pe\alpha\beta \notin M\}$. Then M* is an interpretation of S3 that satisfies S1. If $R\alpha\beta$ and $P\gamma\delta\alpha$ are in M*, then $Pe\alpha\beta$ and $P\gamma\delta\alpha$ must be in M and, since $\overline{P}ezu\ \overline{P}xyz\ Pxyu$ is a theorem of S1, it follows that $P\gamma\delta\beta$ is in M and hence in M*. Thus M* satisfies B14. A similar argument shows that M* satisfies B9 as well, hence M* is a model of S3 and must satisfy D. Let $M^* \cap D = E$. Then $M \cap D\theta = E\theta \neq \emptyset$, contrary to the hypothesis that M condemned Dθ. It remains to show that θ maps only elements of V(S3) into V(S1): Since S3 $\models$ $\overline{P}exy\ Rxy$ and S3 $\models$ $\overline{R}xy\ Pexy$, it follows that S3,Cθ $\models$ C for any $C \in W(S3)$. If S1 $\models$ Cθ, then S3 $\models$ Cθ, since S1 $\subseteq$ S3. Hence if S1 $\models$ Cθ, S3 $\models$ C.

To show that S3 is adequate for A1-8, let μ be the identity mapping on clauses and first note that S3 $\models$ Ai for i - 1,...,8. (Two of the examples in the section on proof search efficiency contain proofs that S3 $\vdash$ B21-22. Proofs that S3,B21-22 $\vdash$ A1-8 are given in Appendix A.) Hence if $C \in V(A1\text{-}8)$, then $C\mu = C \in V(S3)$. Now suppose that $C \in W(A1\text{-}8) - V(A1\text{-}8)$. Then there must be a model M of A1-8 over H_{A1-8} that condemns C. Let $M^* = M \cup \{P\alpha\beta\gamma | Rf(\alpha\beta)\gamma \in M\} \cup \{\overline{P}\alpha\beta\gamma | Rf(\alpha\beta)\gamma \notin M\}$. Now A1-8 $\cup$ {$\overline{R}f(xy)z$ Pxyz, $\overline{P}xyz$ Rf(xy)z} $\models$ S3. Hence M* must be a model of S3. Since M* condemns C, $C\mu = C$ cannot be in V(S3).

In general, if S' is adequate for S and S is adequate for T, it follows that S' is adequate for T, since if μ' is the mapping of S into S' and μ the mapping of T into S, $\mu\mu'$ will do for the mapping of T into S'. In particular, since S1 is adequate for S3 and S3 adequate for A1-8, S1 must be adequate for A1-8 and in turn for any theories for which A1-8 may be adequate. One might wish to note that A1-8 is not a non-creative extension of S1.

Proof-search Efficiency

In the literature on automatic theorem-proving, considerable attention has been devoted to selection of an efficient proof search algorithm. [1], [2], [4], [6], [8], [10]. Nevertheless, at the present state of the art, the ease--even the feasibility--of automatic theorem proving in a given theory (e.g., group theory) is vitally affected by the choice of axioms and representations of theorem-candidates for the study of the theory. Indeed, the choice can be so important that it is difficult to find examples to compare search times for systems such as A1-8 with more efficient (with a given proof search algorithm) systems such as S3. With an unfortunate choice of axioms, the running times for quite simple theorems can be prohibitively long to obtain a numerical comparison.

The figures for running times given in this section are for the PG1-PG5 series of theorem-proving programs developed at Argonne a number of years ago and more fully described in [8], [9], and [10]. PG5 has been singled out for use in most of the cases run because it provides the fairest basis for comparing diverse axiom systems, even though it is slower in some cases than some others in the PG1-PG5 family. In order to obtain numerical comparisons for efficiency of axiom systems in ordinary first-order theories with no special treatment of equality, the demodulation apparatus of PG5 was disabled. It proved to be infeasible to obtain, with the programs available, conclusive efficiency comparisons for first-order theories with equality, due to incompleteness difficulties introduced by the treatment of demodulation in those programs having special treatment of equality.

First we shall consider the example of proving that in a group a right inverse exists. As is usually the case in automatic theorem proving, the procedure is to deny the existence of a right inverse and proceed to refute the denial by appeal to the axioms. In the vocabulary of A1-8, the denial is $\overline{R}$f(ay)e; for S3 it is $\overline{P}$aye. PG5 obtained a refutation from S3 in less than one second, but could obtain no refutation from A1-8 in the 288 seconds allowed for running the case. Some insight into the difficulties may perhaps be obtained by examining refutations in the two systems. One refutation from A1-8 is as follows:

1.	$\overline{R}$xy Ryx	A1-sym
2.	$\overline{R}$xy $\overline{R}$yz Rxz	A2-trans
3.	Rf(f(xy)z)f(xf(yz))	A3-assoc
4.	$\overline{R}$xz $\overline{R}$yu Rf(xy)f(zu)	A5-f-subst
5.	Rf(ex)x	A6-1. ident
6.	Rf(g(x)x)e	A7-1. inv
7.	Rxx	A8-reflx
8.	$\overline{R}$f(ay)e	denial
9.	Rf(xf(yz))f(f(xy)z)	$A3\text{-}A1_1$
10.	$\overline{R}$xz Rf(xf(ey))f(zy)	$A6\text{-}A5_2$
11.	Rf(xf(ey))f(xy)	$10_1\text{-}A8$
12.	$\overline{R}$wf(zf(ex)) Rwf(zx)	$11\text{-}A2_2$
13.	Rf(f(ze)x)f(zx)	$A3\text{-}12_1$
14.	$\overline{R}$wf(g(x)x) Rf(we)	$A7\text{-}A2_1$
15.	Rf(f(g(x)e)x)e	$13\text{-}14_1$
16.	$\overline{R}$f(ax)y $\overline{R}$ye	$8\text{-}A2_3$

17.	$\overline{R}f(ax)f(f(g(y)e)y)$	$15\text{-}16_2$
18.	$\overline{R}af(g(y)e\ \overline{R}xy$	$17\text{-}A5_3$
19.	$\overline{R}af(g(x)e$	$A8\text{-}18_2$
20.	$\overline{R}yu\ Rf(f(g(x)x)y)f(eu)$	$A7\text{-}A5_1$
21.	$Rf(f(g(x)x)y)f(ey)$	$A8\text{-}20_1$
22.	$\overline{R}wf(f(g(x)x)y)\ Rwf(ey)$	$21\text{-}A2_2$
23.	$Rf(g(x)f(xz))f(ez)$	$9\text{-}22_1$
24.	$\overline{R}wf(ez)\ Rwz$	$A6\text{-}A2_2$
25.	$Rf(g(x)f(xz))z$	$23\text{-}24_1$
26.	$Rzf(g(x)f(xz))$	$25\text{-}A1_1$
27.	$\overline{R}f(g(x)f(xz))w\ Rzw$	$26\text{-}A2_1$
28.	$\overline{R}f(g(x)f(xa))f(g(y)e)$	$19\text{-}27_2$
29.	$\overline{R}g(x)g(y)\ \overline{R}f(xa)e$	$28\text{-}A5_3$
30.	$\overline{R}f(ya)e$	$A8\text{-}29_1$
31.	$\square$	A7-30

Contrast that refutation with the following one obtained from S3:

1.	$Pexx$	B6-1.ident
2.	$Pg(x)xe$	B7-1.inverse
3.	$\overline{P}xyu\ \overline{P}yzt\ \overline{P}uzw\ Pxtw$	B11
4.	$\overline{P}xyu\ \overline{P}yzt\ \overline{P}xtw\ Puzw$	B12
5.	$\overline{P}aye$	denial
6.	$\overline{P}xya\ \overline{P}yzt\ \overline{P}xte$	$5\text{-}4_4$
7.	$\overline{P}xea\ \overline{P}xte$	$1\text{-}6_2$
8.	$\overline{P}g(t)ea$	$2\text{-}7_2$
9.	$\overline{P}g(t)yu\ \overline{P}yze\ \overline{P}uza$	$8\text{-}3_4$

10.	$\overline{P}$g(t)ye $\overline{P}$yae	$1\text{-}9_3$
11.	$\overline{P}$yae	$2\text{-}10_1$
12.	□	11-2

For further contrast, we cite the following proof from A1-8 using the special equality mechanisms of paramodulation. This proof suggests that A1-8 (more precisely the subset {A3,A6,A7,A8}) is probably a better choice than S3 when such special mechanisms for equality are incorporated into the theorem-proving program.

1.	Rf(f(xy)z)f(xf(yz))	A3
2.	Rf(ex)x	A6
3.	Rf(g(x)x)e	A7
4.	$\overline{R}$f(ay)e	denial of theorem
5.	Rf(f(xg(z))z)f(xe)	A7(f(g(x)x))-A3(f(yz))
6.	Rf(f(ez)f(g(g(z))e)	A7(f(g(x)x))-5(f(xg(z))
7.	Rzf(g(g(z))e)	A6(f(ex))-6(f(ez))
8.	Rf(f(xe)z)f(xz)	A6(f(ex))-A3(f(yz))
9.	Rf(f(g(z)e)z)e	A7(f(g(x)x))-8(f(xz))
10.	Rf(zg(z)e)	7(f(g(g(z)e))-9(f(g(z)e))
11.	□	10-4

Similar results are obtained when the two systems S3 and A1-8 are used to refute the denial that in a group, the left identity element is also a right identity. A refutation from A1-A8 looks much like that for right inverse. No refutation is obtained by PG5 from this set after

288 seconds, while less than one second is required for a refutation from S3. This is not surprising since short refutations such as the following are available from S3.

1.	Pexx	B6
2.	Pg(x)xe	B7
3.	$\overline{P}xyu\ \overline{P}yzt\ \overline{P}uzw\ Pxtw$	B11
4.	$\overline{P}xyu\ \overline{P}yzt\ \overline{P}xtw\ Puzw$	B12
5.	$\overline{P}aea$	denial
6.	$\overline{P}xya\ \overline{P}yet\ \overline{P}xta$	$5\text{-}B12_4$
7.	$\overline{P}xea$	$B6\text{-}6_2$
8.	$\overline{P}xyu\ \overline{P}yze\ \overline{P}uza$	$7\text{-}B11_4$
9.	$\overline{P}xye\ \overline{P}yae$	$B6\text{-}8_3$
10.	$\overline{P}yae$	$B7\text{-}9_1$
11.	□	10-B7

If we are correct in our conjecture that it is advantageous to use additional free variables and, if necessary, additional literals in order to avoid long terms as arguments, one would expect that adding, say, A11 and A12 (or replacing A3 by A11-12) might improve the performance of that set. PG5 does in fact get a proof of right identity from A1-8,11-12 in less than two seconds, while it failed to find a proof from A1-8 alone in 288 seconds.

If we go to slightly more difficult (to prove from the basic axioms) theorems such as that if in a group the square of every element is the identity then the group is commutative, we find that even S3 is sorely taxed. This is one of a large class of theorems for which proof-search efficiency is greatly improved by the addition of the logically dependent axioms B21-22 for right identity and right inverse. This phenomenon--that inclusion of dependent axioms does not always detract from proof search efficiency but may be a positive benefit, possibly even a necessity--is one of the more important insights into axiom selection, at least with the type of search algorithms we employ in our programs. The denial of the theorem above is $Rf(xx)e \wedge Rf(ab)c \wedge \overline{R}f(ba)c$ for A1-8 and $Pxxe \wedge Pabc \wedge \overline{P}bac$ for S3. The proof from A1-8,A21-22 is again quite tedious:

1.	$\overline{R}xy\ Ryx$	A1
2.	$\overline{R}xy\ \overline{R}yz\ Rxz$	A2
3.	$Rf(f(xy)z)f(xf(yz))$	A3
4.	$\overline{R}xz\ \overline{R}yu\ Rf(xy)f(zu)$	A5
5.	$Rf(ex)x$	A6
6.	Rxx	A8
7.	$Rf(xe)x$	A21
8.	$Rf(xx)e$	denial of theorem
9.	$Rf(ab)c$	" " "
10.	$\overline{R}f(ba)c$	" " "
11.	$Ref(xx)$	$8\text{-}A1_1$
12.	$\overline{R}zu\ Rf(xz)f(xu)$	$A8\text{-}A5_1$
13.	$Rf(ue)f(uf(xx))$	$11\text{-}12_1$
14.	$Rf(xf(yz))f(f(xy)z)$	$A3\text{-}A1_1$
15.	$\overline{R}f(uf(xx))z\ Rf(ue)z$	$13\text{-}A2_1$
16.	$Rf(ue)f(f(ux)x)$	$14\text{-}15_1$

17.	Ruf(ue)	$A21\text{-}A1_1$
18.	$\overline{R}f(ue)z$ Ruz	$17\text{-}A2_1$
19.	Ruf(f(ux)x)	$16\text{-}18_1$
20.	Rf(xu)f(xf(f(uw)w))	$19\text{-}12_1$
21.	$\overline{R}uf(xf(yz))$ Ruf(f(xy)z)	$14\text{-}A2_2$
22.	Rf(xu)f(f(xf(uw))w)	$20\text{-}21_1$
23.	$\overline{R}zu$ Rf(zy)f(uy)	$A8\text{-}A5_2$
24.	Rf(f(zz)yf(ey)	$8\text{-}23_1$
25.	$\overline{R}xf(f(zz)y)$ Rxf(ey)	$24\text{-}A2_2$
26.	Rf(f(uw)u)f(ew)	$22\text{-}25_1$
27.	$\overline{R}xf(ez)$ Rxz	$A6\text{-}A2_2$
28.	Rf(f(uw)u)w	$26\text{-}27_1$
29.	$\overline{R}xf(ab)$ Rxc	$9\text{-}A2_2$
30.	$\overline{R}f(ba)f(ab)$	$10\text{-}29_2$
31.	$\overline{R}f(ab)f(ba)$	$30\text{-}A1_2$
32.	$\overline{R}f(ab)y$ $\overline{R}yf(ba)$	$31\text{-}A2_3$
33.	$\overline{R}f(f(f(ab)z)z)f(ba)$	$19\text{-}32_1$
34.	$\overline{R}f(f(ab)a)b$	$33\text{-}23_2$
35.	$\square$	34-28

The proof from S3 $\cup$ {B21,B22} is, as before, shorter (still shorter proofs than the one given below have been obtained by the computer):

1.	Pexx	B6
2.	Pxex	B21
3.	Pxg(x)e	B22
4.	$\overline{P}xyu\ \overline{P}yzt\ \overline{P}uzw\ Pxtw$	B11
5.	$\overline{P}xyu\ \overline{P}yzt\ \overline{P}xtw\ Puzw$	B12
6.	Pxxe	denial
7.	Pabc	"
8.	$\overline{P}bac$	"
9.	$\overline{P}xye\ \overline{P}ywt\ Pytw$	$1\text{-}4_3$
10.	$\overline{P}ywt\ Pytw$	$6\text{-}9_1$
11.	Pacb	$7\text{-}10_1$
12.	$\overline{P}yzt\ \overline{P}ezw\ Pytw$	$6\text{-}4_1$
13.	$\overline{P}bza\ \overline{P}ezc$	$8\text{-}12_3$
14.	$\overline{P}wyu\ \overline{P}yze\ Puzw$	$2\text{-}5_3$
15.	$\overline{P}g(w)ze\ Pezw$	$3\text{-}14_1$
16.	Peg(w)w	$6\text{-}15_1$
17.	$\overline{P}bg(c)a$	$16\text{-}13_2$
18.	$\overline{P}wyu\ Pug(y)w$	$3\text{-}14_2$
19.	Pbg(c)a	$11\text{-}18_1$
20.	□	17-19

References

1. Kowalski, R., and Hayes, P. "Semantic trees in automatic theorem proving," *Machine Intelligence IV*, ed. by D. Michie and B. Meltzer (1969).

2. Luckham, D. "Some tree-paring strategies for theorem-proving." *Machine Intelligence III*, ed. by D. Michie, Edinburgh Univ. Press, Edinburgh (1968).

3. Robinson, A. *Model theory and the mathematics of algebra.* North Holland, Amsterdam (1963).

4. Robinson, G., and Wos, L. "Paramodulation and theorem-proving in first-order theories with equality." *Machine Intelligence IV*, ed. by D. Michie and B. Meltzer (1969).

5. Robinson, J. "A machine-oriented logic based on the resolution principle." *J.ACM* 12 (1965), pp. 23-41.

6. Slagle, J. "Automatic theorem proving with renumerable and semantic resolution." *J. ACM* 14 (1967), pp. 687-697.

7. Suppes, P. *Introduction to Logic*, Prentice Hall, New York (1957).

8. Wos, L., Carson, D., and Robinson, G. "The unit preference strategy in theorem proving." *AFIPS Conf. Proc.* 26, Spartan Books, Washington D. C. (1964), pp. 615-621.

9. Wos, L., Robinson, G., and Carson, D. "Efficiency and completeness of the set of support strategy in theorem proving." *J. ACM* 12 (1965), pp. 536-541.

10. Wos, L., Robinson, G., Carson, D., and Shalla, L. "The concept of demodulation in theorem proving." *J. ACM* 14 (1967), pp. 698-709.

Appendix A

Proof that S3,B21-22 A1-8:

	1.	$\overline{P}exu\ Rxu$	$B6-B9_1$	(resolution of B6 against first literal of B9)
A8:	2.	Rxx	$B6-1_1$	(resolution of B6 against first literal of step 1)
	3.	$\overline{R}zu\ \overline{P}xyz\ \overline{P}xyw\ Ruw$	$B14_3-B9_1$	(resolution of third literal of B14 against first literal of B9)
	4.	$\overline{R}zu\ \overline{P}xyz\ Ruz$	3_{2-3}	(factoring step 3 on second and third literals)
A1:	5.	$\overline{R}zu\ Ruz$	$B6-4_2$	
	6.	$\overline{R}yu\ Peyu$	$B6-B14_2$	
	7.	$\overline{R}yz\ \overline{R}zu\ Peyu$	$B14_2-6_2$	
	8.	$\overline{R}yz\ \overline{R}zu\ \overline{P}eyw\ Rwu$	$B9_2-7_3$	
A2:	9.	$\overline{R}yz\ \overline{R}zu\ Ryu$	$B6-8_3$	
	10.	$\overline{P}xyu\ Rf(xy)u$	$B13-B9_1$	
	11.	$\overline{P}yzt\ \overline{P}xtw\ Pf(xy)zw$	$B13-B12_1$	
	12.	$\overline{P}xf(yz)w\ Pf(xy)zw$	$B13-11_1$	
	13.	$Pf(xy)zf(xf(yz))$	$B13-12_1$	
A3:	14.	$Rf(f(xy)z)f(xf(yz))$	$13-10_1$	
	15.	$\overline{R}yu\ \overline{P}yzt\ \overline{P}etw\ Puzw$	$B12_1-6_2$	
	16.	$\overline{R}yu\ \overline{P}yzt\ Puzt$	$B6-15_3$	
	17.	$\overline{R}xu\ Pxeu$	$B21-B14_2$	
	18.	$\overline{R}xu\ \overline{P}yxz\ \overline{P}zew\ Pyuw$	$B11_2-17_2$	
	19.	$\overline{R}xu\ \overline{P}yxz\ Pyuz$	$B21-18_3$	
	20.	$\overline{R}uy\ \overline{P}yzt\ Puzt$	5_2-16_1	
	21.	$\overline{R}ux\ \overline{P}yxz\ Pyuz$	5_2-19_1	
	22.	$\overline{R}ux\ \overline{P}yxz\ \overline{R}wy\ Pwuz$	21_3-20_2	
	23.	$\overline{R}wy\ \overline{R}ux\ Pwuf(yx)$	$B13-22_4$	

A5:	24.	$\overline{R}wy\ \overline{R}ux\ Pf(wu)f(yx)$	23_3-10_1
	25.	$\overline{P}xzt\ \overline{P}ezw\ Pg(x)tw$	B7-$B11_1$
	26.	$\overline{P}xzt\ Pg(x)tz$	B6-25_2
	27.	$\overline{R}tx\ \overline{P}tzu\ Pg(x)uz$	26_1-16_3
	28.	$\overline{R}tx\ Pg(x)eg(t)$	B22-27_2
	29.	$\overline{P}yeu\ Ruy$	B21-$B9_2$
A4:	30.	$\overline{R}tx\ Rg(t)g(x)$	29_1-28_2
A6:	31.	$Rf(ex)x$	B6-10_1
A7:	32.	$Rf(g(x)x)e$	B7-10_1

CONSTRUCTIVE VALIDITY

DANA SCOTT

This paper is a preliminary report on work in progress and is an expanded and revised version of the lecture given at the conference. The author is indebted to N.G. de BRUIJN, N. GOODMAN, G. KREISEL and A.S. TROELSTRA for many kinds of help, information and advice as well as stimulation. In particular KREISEL has been very patient over the years in repeating time after time points not taken in and in offering extended criticism of faulty attempts at understanding what he calls "non-set-theoretic" foundations. The author is also indebted to D. LACOMBE for bringing a formal decidability problem to his attention, and to G. KREISEL for discussions on the significance of this problem. (See postscript).

BACKGROUND

A quote from HEYTING (8) will set the stage as well as could be desired:

One of BROUWER'S main theses was that mathematics is not based on logic, but that logic is based on mathematics. This is easily seen to be an immediate consequence of his point of departure. If mathematics consists of mental constructions, then every mathematical theorem is the expression of a result of a successful construction. The proof of the theorem consists in this construction itself, and the steps of the proof are the same as the steps of the mathematical construction. These are intuitively clear mental acts, and not applications of logical laws. Yet an intuitionistic logic has been developed, and thus the question of its significance was raised. The older interpretations by KOLMOGOROFF (as a calculus of problems) and HEYTING (as a calculus of intended constructions) were substantially equivalent. In a later paper HEYTING interprets logical theorems as simply mathematical theorems of extreme generality. There is no essential difference between logical and mathematical theorems, because both sorts of theorems affirm that we have succeeded in performing constructions satisfying certain conditions.

BROUWER based his considerations on a complex philosophical standpoint and a thorough psychologistic view of the nature of mathematics. Our purpose here will be to reexamine the idea of the calculus of constructions. A formalization of this calculus will be presented, and it will be applied to the problem of interpreting logical formulas in a way that, to the author at least, seems to carry out the program outlined by HEYTING above word for word. When this is done it would appear that the psychologism has been reduced to a minimum: one only has to agree that the theory of constructions has intuitive appeal. And one particular advantage of the theory we shall examine is that it has many interpretations of varying degrees of constructivity. Now there will of course remain the questions of whether BROUWER would have considered the theory at all reasonable and of whether some essential part of his idea of mathematics has been lost. But the author feels that until the intuitionists arrive at a greater degree of clarity in formulating their principles, the conclusion must stand that the notion presented here is indistinguishable from the intended meaning on the basis of current practice, of intuitionistic mathematics.(This statement is incorrect; see postscript.)

These remarks <u>do not</u> apply <u>directly</u> to BROUWER'S theory of choice sequences, but the present state of the art (cf.(19) and the objections of MYHILL (17)) indicates that choice sequences are <u>eliminable</u>. Thus, however pleasant they may be in theory (and natural in intuition), one cannot claim for them at the moment any more fundamental rôle in analysis than, say, that of the <u>infinitesimals</u> of (classical) non-standard analysis. For both kinds of analysis these various remarkable reals have properties that aid our understanding through the regularity of their laws, but strictly speaking they are not needed. This situation may very well change in the light of future developments; hence the cautious reader may be reluctant to call the author's theory <u>intuitionism</u>.

The calculus we shall develop here did not occur as a bolt out of the blue but has a long history involving many people. In the first place we have HEYTING'S original work. The author's own contact with the problem came through KREISEL'S formulations in (10) and (11). Subsequently interest was revived in consulting with GOODMAN on the thesis (5) (cf. also (6), and more on this later. In the meantime we had the work by LÄUCHLI (14) and LAWVERE (15) who both provided interpretations that are closely related. Their approach has one serious defect from our point of view: neither of them formalized their theories of functions (constructions) and both of them think rather non-constructively. (I hope they will forgive me for this remark.) Therefore the foundational (as distinct from mathematical) content of their interpretations is not evident. Hopefully the present theory will make it possible to view their results in a new light. LÄUCHLI was motivated by KLEENE'S realizability interpretation (cf.(9)) and considered his notion as an abstract generalization thereof. The exact relation of the present interpretation to realizability is not clear yet. KLEENE'S particular use of recursive functions introduces anomalies (sometimes formally useful!) that make comparison difficult. GOODMAN discusses this in (5), but we shall not be able to do so here.

Getting back to KREISEL, he wanted to formalize the "intended" interpretation in such a way that proofs (in an abstract sense) were objects of the theory of the same status as constructions. This is reasonable from the psychologistic approach which accepts mental acts as objects of mathematical investigation. There were some difficulties in bringing KREISEL'S theory to a precise enough state to allow metamathematical results, and this problem was the point of departure for GOODMAN. He reformulated KREISEL'S theory and obtained several results, but his version was not exactly what KREISEL had wanted. KREISEL felt that in view of decidability of various features of proofs, the functions should be total functions. GOODMAN did not find this requirement convenient because operations on constructions were to be given by general combinators (in the sense of Curry-Church), and these necessarily lead to partial functions. GOODMAN gave a quite neat treatment of a calculus of partial functions, and aside from this divergence carried out KREISEL'S plan in satisfactory detail. It will be noted, however, that neither KREISEL nor GOODMAN gave an analysis of the structure of abstract proofs, and they enter in a (to the author) mysterious way simply to allow certain properties to be decidable.

This was how matters stood at the time the author came to Amsterdam in the fall of 1968. Soon thereafter he met Professor de BRUIJN, who explained to him his language AUTOMATH (cf. the paper of de BRUIJN at this conference). The feature of his language what was of special interest to de BRUIJN was the possibility of writing a computer program for practical proof checking but that will not concern us for the moment. What was highly suggestive to the author was de BRUIJN'S conceptual framework. He had been, of course, personally influenced by BROUWER and wanted to present a suitably constructive notion of proof. He achieved this, not surprisingly, by means of constructions for interpreting the logical notions. He distinguishes between constructions (functions) and categories (certain sets or species) of constructions and places the burden of proof on showing that a given compound construction belongs to the desired category. The particular conventions of language for writing such proofs, which are essential for computer work, need not be discussed here.

As the reader can well imagine, at this point the author made the connection with KREISEL, GOODMAN, LÄUCHLI, and LAWVERE, and he set out to formulate a system of his own. Instead of the natural deduction style of de BRUIJN, it seemed more succinct to use the calculus of sequents employed by GOODMAN for foundational considerations. (This also seems better than the two-valued propositional connectives of KREISEL, since one is only interested in certain implications in any case.) Next the distinction between constructions and species used by de BRUIJN seemed very convenient, though as we shall see this does not require notational distinctions. When one does this one finds that partial functions can be <u>avoided</u> by having each function defined on a "principal" domain and then making function values arbitrary outside this domain (a plan of KREISEL). Next de BRUIJN made good use of <u>cartesian products</u> of species (formation of function spaces) in connection with the <u>universal quantifier</u> - an idea also familiar to LAWVERE and to a certain extent to LÄUCHLI - and the author took this at once. Now dual to products (as LAWVERE knows) are <u>disjoint sums</u> which must be used for the interpretation of the <u>existential quantifier</u> (cf. KREISEL - GOODMAN). These sums were not employed by de BRUIJN, but it would be easy to add them to his system.

Now that we have functions and species and sums and products, we take certain primitive species (a one-and two-element species, and the species of natural numbers) and their implied functions (ordered pairs and definitions by recursion) and combine and recombine them as much as we please obtaining the <u>basis species</u> of constructive mathematics (cf. the discussion in TROELSTRA (20)). These are finally used for the interpretation of logic.

Several points should be noted:

(1) We <u>never</u> have occasion to form species <u>of</u> species. Why? Well since we can form functions of functions of ... of functions of species,

the species of species do not seem to be needed. If we can think of some use for them, the format of the theory will allow for them.

(ii) General species variables (quantification over species) are not allowed, though the effect can be produced by some simple primitives. This is not a defect, because there may be arguments against quantification over arbitrary species.

(iii) We have no abstract proofs only constructions and species of constructions. When the author finally obtained his formalism the proofs-as-objects vanished. May be they should be brought back in, but for the present the author's system seems to be simpler than KREISEL - GODDMAN'S (and to a certain extent, de BRUIJN'S) and to be adequate. Thus it seems more reasonable to try it out first in some detail; only then will one be able to appreciate whether abstract proofs are desirable. (But see postsript.)

(iv) The general combinators are not used. This has the advantage that models are conceptually easier to obtain (total vs. partial functions, as mentioned earlier). Furthermore, one is forced to make explicit all the basic modes of formation, and they are remarkably few.

Let us digress for a moment to discuss the category-theoretic approach of LAWVERE. In category theory we axiomatize a calculus of functions under composition. We do not, however, have (what seems to the author) a convenient axiomatization of which infinitary operations (such as direct product) actually exist. Usually we consider a category as a class and talk about (arbitrary) indexed families of objects. Thus the existence of these families is thrown back to set theory.

If there were an axiomatization of the "category" of "all" categories, this would not be necessary, but in the author's opinion this all-inclusive theory does not yet exist. Even if it did, it would most likely not be a constructive theory. If one likes, one can view the author's theory as an attempt at axiomatizing in a constructive way a theory of both functions and families of sets of functions. Whether this approach could have any effect on category theory is a matter of speculation.

At this point, mention should also be made of TAIT'S paper [18]. He called his work "Constructive Reasoning" and seemed to make a conscious effort not to define validity. He does of course discuss the GÖDEL interpretation, but that is not the same thing. Also he uses in an essential way definitional equality which we have not found necessary, though it is a notion favored by KREISEL. Furthermore, TAIT'S use of combinators leads to a theory of species that does not seem as elementary as the one presented here. The author does, however, agree with TAIT on the introduction of species of trees used to index quite general iterations and will discuss this in detail below.

In summary, then, based on the motivations and contacts just explained, we are going to propose a theory of constructions and species and to show how it applies in making precise the meanings of the logical notions. This theory involves the primitive ideas of sums, products and iterations applied to the finite species to generate the basic species which provide the background for constructive mathematical thought.

LANGUAGE

We shall distinguish as usual between terms and formulas; however, only the terms will be compounded not the formulas. Thus as formulas we have:

$$\sigma \in \tau \quad \text{and} \quad \sigma = \tau$$

where σ and τ are terms. The first means that the (construction) σ belongs to the (species) τ ; while the second is an equation (between constructions or species). There seems to be no need at all to have a two-sorted theory (indeed later it would actually be inconvenient), so we have just one type-free sort of variable (usually, lower-case Roman letters with the Greek letters reserved for metatheory) ranging over both constructions and species, variables are terms.

Among the terms we mention next the constants, namely:

$$\mathbf{0}, \dot{0}, \mathbf{1}, \bar{0}, \bar{1}, \mathbf{2},$$

of these $\mathbf{0}$, $\mathbf{1}$, and $\mathbf{2}$ are thought of as species and the others as (atomic) constructions. (One may guess the membership relations now, but they are made explicit later by axioms.)

Further we have some simple compound terms, namely:

$$\sigma(\tau),\ \sigma_0, \sigma_1, [\sigma \wedge \tau],\ O(\sigma), \sigma^+, \mathbf{\Pi}(\sigma),$$

where σ and τ are previously obtained terms. Explanations of meanings now deserve special sections below.

Finally we have the complex compound terms which involve bound variables:

$$\forall x \varepsilon \alpha[\sigma],\quad \exists x \varepsilon \alpha[\sigma],\quad \mathrm{P} x \varepsilon \alpha[\sigma],\quad \text{and } \mathrm{R} v[\alpha, \beta, \sigma].$$

Here α, β, σ are terms and in place of x and v we may have any other variables. One should not worry now about the use of ε: it could be just another punctuation mark. The reason for the placement of brackets is that our's is more a postfix rather than a prefix notation. This could be modified, but it makes formulas even less beautiful. The reason for writing α, β is that our convention is such that the variable (x or v) is bound only in the σ not in the α or the β. One defines free and bound occurrences of variables in the usual way as well as the notion of rewriting bound variables.

We shall often have to indicate the substitution of a term σ for all free occurences of a variable (x, say) with the implied rewriting of variables free in σ if they occur bound in the context (τ, say) into which we substitute. We use the notation

$$[\sigma/x][\tau]$$

(sometimes without the second pair of brackets) and remark that this is a notation of the metalanguage not the object language.

INFERENCE

Connections between formulas are asserted by sequents

$$\Delta \vdash \delta$$

where Δ is a (finite) sequence of formulas and δ is a single formula. The meaning is clear: the conjunction of the formulas in Δ implies the formula δ. We provide no brackets because this implication is never iterated.

A stock of these assertions is provided later by the axioms; while the theorems are derived from them by these well-known rules:

(I1) $\dfrac{\Delta \vdash \delta}{\Delta, \varepsilon \vdash \delta}$ (Weakening)

(I2) $\dfrac{\Delta, \varepsilon, \varepsilon', \Lambda \vdash \delta}{\Delta, \varepsilon', \varepsilon, \Lambda \vdash \delta}$ (Interchange)

(I3) $\dfrac{\Delta, \varepsilon \vdash \delta \quad \Delta \vdash \varepsilon}{\Delta \vdash \delta}$ (Cut)

(I4) $\dfrac{\Delta \vdash \delta}{\Delta \vdash \delta'}$ (Rewrite)

(I5) $\dfrac{\Delta \vdash \delta}{[\sigma/x]\Delta \vdash [\sigma/x]\delta}$ (Substitution)

where in the rewrite rule, δ' results from δ by rewriting a bound variable. (It may be possible that (I4) follows from substitution if we made the substitution conventions really precise, but never mind.)

The author does not know whether it is true - or even interesting to suppose that there is a "cut-free" formulation of the theory. This question might be related to some decision problems mentioned below.

EQUALITY - Two axioms are required:

(E1) $\vdash x = x$

(E2) $x = y, \delta \vdash [y/x]\delta$

these hardly need explanation.

We note these obvious consequences:

$$\delta \vdash \delta$$
$$x = y, x = z \vdash y = z$$
$$x = y \vdash y = x \quad .$$

It is possible that equality could be eliminated from the system, but it does not seem pleasant to do so.

FUNCTIONS

We mean by $f(x)$ the ordinary function value f of x . In as much as functions can take functions as values, we can write $f(x)(y)$ for functions of two arguments, and similarly for more arguments. All our functions are total, so that $f(x)$ always means something even if x

is outside the principal domain of f. In that case, we would, if we so desired, let $f(x) = f$, but we shall formulate no axiom to that effect leaving the matter open.

For the most part a theory of functions is quite uninteresting unless there is some method for introducing new functions by (explicit) definition. We provide such definitions through functional abstraction. Thus if σ is a term (with the variable x free in σ, say) and if a is a given species, then we can think of the function f defined on a with value σ for $x \varepsilon a$. Our notation for this function is:

$$f = \forall x \varepsilon a[\sigma] .$$

Most people will consider the author slightly mad to use the universal quantifier for functional abstraction. Nevertheless, there is method in his madness as will be clear in the next section. In the meantime, the reader may rub out the $\forall$ and replace if by λ, if that makes him happier. The idea of functional abstraction is formalized in the axiom of conversion:

(F1) $$f = \forall x \varepsilon a[\sigma],\ x \varepsilon a \vdash f(x) = \sigma.$$

Note the variations of the axiom that can be obtained by substitution.

To the non-constructive mind it would seem reasonable to adjoin at this point the rule of extensionality;

$$\frac{\Delta,\ x \varepsilon a \vdash \sigma = \tau}{\Delta \vdash \forall x \varepsilon a[\sigma] = \forall x \varepsilon a[\tau]}$$

where x is not free in Δ. Exactly why this is unreasonable the author cannot argue convincingly at the moment. However, to leave open possible "intensional" interpretations of the axioms (the same functions may be given by different GÖDEL numbers, say), it seems better to avoid it. In any case it was not required for the interpretation of logic.

We make one apparently harmless concession to extensionality though:

(F2) $$f = \forall x \varepsilon a[\sigma] \vdash f = \forall x \varepsilon a[f(x)] .$$

This may not really be needed, but the equation on the right is a way of saying that the principal domain of f is a. (We also considered having an operator Df = a for computing domains, but dropped the idea as unnecessary.)

So much for single functions, we must now turn to the consideration of species of functions.

PRODUCTS

Familiar from set theory, topology, and algebra is the cartesian (direct) product. Familiar too is its fundamental rôle, and so it will be here. Given species $\sigma(x)$ indexed by $x \varepsilon a$, we consider all functions f defined on a such that $f(x) \varepsilon \sigma(x)$ for all $x \varepsilon a$. These form a (basic) species, whose existence we wish to postulate. First being influenced by ordinary mathematics, we might call it:

$$\mathsf{X} x \varepsilon a [\sigma(x)].$$

But let us stop to think a moment. We have distinguished between species and functions, because we must give the functions a special place. (In mathematics the idea of function really is more primitive than most other notions.) In particular there is absolutely no reason to identify a function with a set of ordered pairs as is usual in (pure) set theory. For one thing such an identification is not particularly constructive; for another, our species are rather more restricted that those allowed in set theory. So another plan may be considered.

Let us reason as follows: for the moment functions and species are separated. Maybe an identification can be reestablished that is even more convenient than the usual one. (An identification is a simplification - hopefully - that avoids proliferation of entities.) In our case the product $\mathsf{X} x \varepsilon a [\sigma(x)]$ is completely determined by the function $\forall x \varepsilon a [\sigma(x)]$. Conversely, assuming as we do that no species is known to be empty, then the product also determines the function (this point is not too essential). Hence, no one can stop us from making the identification:

$$\mathsf{X} x \varepsilon a [\sigma(x)] = \forall x \varepsilon a [\sigma(x)],$$

and we therefore drop the X notation. Of course, it remains to be seen whether the identification (which was partly suggested by Professor de BRUIJN'S style) is actually useful.

Now that we have the idea of products as functions, we can formulate the abvious axioms. In the first place we have an analogue to (F1):

(P1) $$f \varepsilon \forall x \varepsilon a[\sigma],\ x \varepsilon a \vdash f(x) \varepsilon \sigma \ .$$

Next we also take an analogue to (F2):

(P2) $$f \varepsilon \forall x \varepsilon a[\sigma] \vdash f = \forall x \varepsilon a[f(x)] \ .$$

Finally we must assume what would be an analogue to the rule of extensionality:

(P3) $$\frac{\Delta,\ x \varepsilon a \vdash \sigma \varepsilon \tau}{\Delta \vdash \forall x \varepsilon a[\sigma] \varepsilon \forall x \varepsilon a[\tau]} \ ,$$

where x is not free in Δ. Axioms (P1) and (P2) tell us that the elements of a product have the proper character; while (P3) expresses the fact that any function of the proper kind must belong. In contradistinction to extensionality, this rule is harmless: even though there may be several "copies" of the same function (given by different definitions, say), we can obviously demand that all the copies belong to the product. Note that (P3) is very much like the rule of universal generalization, and this apt analogy will be exploited later.

Once we can form products, they can be specialized to what are usually called powers. For reasons that will eventually become apparent, we use this definition (which may be considered as a new axiom by prefixing the $\vdash$):

DEFINITION

$$[a \to b] = \forall x \varepsilon a[b] \ .$$

As a function $[a \to b]$ is the constant function on a, as a species the use of the notation $[a \to b]$ does not differ too much from the use of the usual category-theoretic notation $f : a \to b$, but we have to write $f \varepsilon [a \to b]$. We also find that $\to$ does indeed behave like (intuitionistic) implication, but before we discuss this in detail the reader might try this theorem as an exercise:

$$x \varepsilon a \vdash \forall f \varepsilon \forall x \varepsilon a[\sigma][f(x)] \varepsilon [\forall x \varepsilon a[\sigma] \to \sigma] \ .$$

The assertion results from (P1) by the rule (P3), and the part underlined should be considered as a whole. What is interesting is to the right of the principal ε, an expression reminding one of the logical axiom of universal instantiation.

Here is another simple exercise:

$$\vdash \forall x \varepsilon a [\forall y \varepsilon b [x]] \ \varepsilon \ [a \rightarrow [b \rightarrow a]] \ .$$

Is this not also suggestive? Would you care to fill in the blank in the theorem:

$$\vdash \underline{\qquad\qquad} \ \varepsilon \ [[a \rightarrow [b \rightarrow c]] \rightarrow [[a \rightarrow b] \rightarrow [a \rightarrow c]]] \ .$$

(The format is deceptive because considerably more space for writing the answer is required than is indicated !) Once you have the idea such examples may be multiplied at will. (This was clear to LÄUCHLI and LAWVERE, for instance.)

SUMS

Dual to products are (disjoint) sums. At this stage we cannot expect any new, clever notational inovations because our previous identifications have exhausted the raw material provided by the functions. So if $\sigma(x)$ are species for $x \in a$, the disjoint sum (union) of these species will be denoted by a new symbol:

$$\exists x \varepsilon a [\sigma(x)].$$

(Note that we can usually omit the "of x" by considering x a free variable of σ ; this is a more formal approach but a little harder to read.) In ordinary set theory the disjoint union is identified with:

$$\bigcup_{x \in a} (\{x\} \times \sigma(x)),$$

but in our theory the ordered pairs cannot be combined in such an arbitrary fashion. For one reason, we are trying to keep our species disjoint (basic species are very much like - a generalization of - the types in the theory of types), and so the same ordered pair cannot belong to distinct species. For another reason, the reduction of disjoint union to ordinary (cumulative) union is not constructive (information is lost in a cumulative union). These considerations thus lead to a related but independent analysis of the notion.

What we seem to have to do next is to provide a separate notion of ordered pair for each district sum. This is just a bit clumsy, but the author could not find a simpler device. Thus the pairing function appropriate to the sum $\exists x \varepsilon a[\sigma(x)]$ is called:

$$\mathbf{P}x\varepsilon a[\sigma(x)] \ ,$$

and we can state the first two axioms governing this pairing function:

(S1) $f = \mathrm{P}x\varepsilon a[\sigma],\ x\varepsilon a,\ y\varepsilon\sigma \vdash f(x)(y)\varepsilon \exists x\varepsilon a[\sigma]$

(S2) $f = \mathrm{P}x\varepsilon a[\sigma] \vdash f = \forall x\varepsilon a[\forall y\varepsilon\sigma[f(x)(y)]]$.

Obviously, the import of these axioms is that the pairing function <u>is</u> a function of the correct type with values in the desired species. This does not yet characterize the values as ordered pairs, however.

It seemed necessary to provide a distinct pairing for distinct sums, because the coordinates of the pair do <u>not</u> determine the context of their occurence (at most the a and the <u>one</u> $\sigma(x)$ is determined by $x\varepsilon a$ and $y\varepsilon\sigma(x)$). On the other hand, we are quite free to assume that the resulting <u>pair</u> $(\mathrm{P}x\varepsilon a[\sigma](x)(y))$ does indeed completely determine not only the coordinates but the whole sum. Hence we can now simplify matters by introducing <u>universal inverse pairing operations</u> that require no special mention of context. The notation is given by the (bold-face) subscripts $\mathbf{0}$ and $\mathbf{1}$, and we have these straight-forward axioms:

(S3) $f = \mathbf{P}x\varepsilon a[\sigma], x\varepsilon a, y\varepsilon\sigma \vdash f(x)(y)_{\mathbf{0}} = x,$

(S4) $f = \mathbf{P}x\varepsilon a[\sigma], x\varepsilon a, y\varepsilon\sigma \vdash f(x)(y)_{\mathbf{1}} = y,$

(S5) $f = \mathbf{P}x\varepsilon a[\sigma], z\varepsilon\exists x\varepsilon a[\sigma] \vdash f(z_{\mathbf{0}})(z_{\mathbf{1}}) = z,$

(S6) $z\varepsilon\exists x\varepsilon a[\sigma] \vdash z_{\mathbf{0}}\varepsilon a,$

(S7) $z\varepsilon\exists x\varepsilon a[\sigma], x = z_{\mathbf{0}} \vdash z_{\mathbf{1}}\varepsilon\sigma$.

It takes several statements, but all we have said here is that the elements of $\exists x \in a[\sigma]$ really do correspond to ordered pairs with well-behaved coordinates.

If the reader wishes a simple exercise, he may prove:

$$\vdash \underline{\mathrm{P}x\varepsilon a[\sigma]}\,\varepsilon\forall x\varepsilon a[[\sigma \rightarrow \exists x\varepsilon a[\sigma]]] \ .$$

GENERATORS

As of now our theory is empty, because we have not yet introduced any species, and so there are no domains on which to define functions. This gap we now fill by providing the finite species from which all the other basic species will be generated. In view of the products and sums, we will only need the first few of these species: $\mathbb{0}$, $\mathbb{1}$, and $\mathbb{2}$. These seem to be independent, and the author doubts that any further simplification is possible.

The species $\mathbb{0}$ is to be empty, but we shall not assume the axiom

$$x \in \mathbb{0} \vdash \sigma$$

at the present. The author cannot put his finger on the precise reason, but somehow this assumption is too strong (there is some connection here with extensionality). Instead we merely remain silent: no axiom or theorem will ever produce an element of $\mathbb{0}$. (This cautiousness is actually unnecessary.)

Hence, if we find one in a hypothetical proof, we know something is absurd or trivial. So much for $\mathbb{0}$.

The species $\mathbb{1}$ is to be the one-element species. It has a much more "positive" character than , and so its axioms are clear:

(G1) $\vdash \mathring{0} \in \mathbb{1}$

(G2) $x \in \mathbb{1} \vdash x = \mathring{0}$.

Thus $\mathring{0}$ is the only element of $\mathbb{1}$. (There is absolutely no real saving in mixing types by the identification $\mathbb{0} = \mathring{0}$, as we do in set theory.) Note that we can easily define functions on $\mathbb{1}$, since such a function has only one value (y, say) and can be defined as:

$$\forall x \in \mathbb{1}[y] = [\mathbb{1} \to y] .$$

It is quite possible that it is sensible to generalize (G2) to the following instance of extensionality.

(G2) $f = \forall x \in \mathbb{1}[f(x)] \vdash f = [\mathbb{1} \to f(0)]$.

This is generalization because (G2) can be derived by substituting

$\forall x \in \mathbf{1}[x]$ for f. (The reader may carry out this simple exercise.) Extensionality of function on finite species seems unexceptional. (The reader may also use (G2') to derive the rule:

$$\frac{\Delta,\ x \in \mathbf{1} \vdash f(x) = g(x)}{\Delta \vdash \forall x \in \mathbf{1}[f(x)] = \forall x \in \mathbf{1}[g(x)]}$$

where x is not free in Δ).

The species $\mathbf{2}$ is to be the two-element species. Since it must be kept (potentially) disjoint with $\mathbf{1}$, it has its own elements $\bar{0}$ and $\bar{1}$. Thus:

(G3) $\vdash \bar{0} \in \mathbf{2}$,

(G4) $\vdash \bar{1} \in \mathbf{2}$.

To say that these are the only elements, we must resort to a rule:

(G5) $$\frac{x = \bar{0},\ \Delta \vdash \delta \qquad x = \bar{1},\ \Delta \vdash \delta}{x \in \mathbf{2},\ \Delta \vdash \delta} .$$

So much is clear; what is not yet clear is how to obtain functions on $\mathbf{2}$. Constant or identity functions are already at hand, but the function that interchanges $\bar{0}$ and $\bar{1}$ is not. In general we must obtain with the aid of a new primitive operation the arbitrarily defined function on $\mathbf{2}$. Suppose its values are to be a and b corresponding to $\bar{0}$ and $\bar{1}$, then we call this function $[a \wedge b]$ and assume:

(G6) $\vdash [a \wedge b](\bar{0}) = a$,

(G7) $\vdash [a \wedge b](\bar{1}) = b$,

(G8) $\vdash [a \wedge b] = \forall x \in \mathbf{2}[[a \wedge b](x)]$,

(G9) $f = \forall x \in \mathbf{2}[f(x)] \vdash f = [f(\bar{0}) \wedge f(\bar{1})]$.

This last is extensionality for the species $\mathbf{2}$.

Here is a useful exercise: prove the following:

$$[x \wedge y] \in [a \wedge b] \vdash x \in a,$$
$$[x \wedge y] \in [a \wedge b] \vdash y \in b,$$
$$x \in a,\ y \in b \vdash [x \wedge y] \in [a \wedge b] .$$

Thus the reader can see that not only does $[a \wedge b]$ play the rôle of an ordered pair, but it is also the same as the finite cartesian product we usually call $[a \wedge b]$. (But the ordered pairs are necessarily distinct from the pairs we needed for sums; sad but true.) If he likes, the reader may also fill in the blanks in these theorems:

$$\vdash \_\_\_\_\_\_\_\_\_\_ \in [[a \wedge b] \to a],$$
$$\vdash \_\_\_\_\_\_\_\_\_\_ \in [[a \wedge b] \to b],$$
$$\vdash \_\_\_\_\_\_\_\_\_\_ \in [a \to [b \to [a \wedge b]]].$$

Suggestive?

Now that we have finite products we should also try to obtain finite sums. Fortunately, these can naturally be identified with combinations already available. By analogy with (G8) we have this definition instead of a new axiom:

DEFINITION

$$[a \vee b] = \exists x \in \mathbf{2}[[a \wedge b](x)].$$

We find here a new quality of the expression $[a \wedge b](x)$. In fact, this is what is usually called in computer science now the <u>conditional expression</u>. If $x = \bar{0}$, the value is a; otherwise if $x = 1$, the value is b. Ready for your exercises? Fill in the blanks, please:

$$\vdash \_\_\_\_\_\_\_\_\_\_ \in [a \to [a \vee b]],$$
$$\vdash \_\_\_\_\_\_\_\_\_\_ \in [b \to [a \vee b]],$$
$$\vdash \_\_\_\_\_\_\_\_\_\_ \in [[a \to c] \to [[b \to c] \to [[a \vee b] \to c]].$$

After so many of these exercise, surely the reader is getting the point and can begin to guess at a general statement.

It would be possible to identify $\mathbf{2}$ with $[\mathbf{1} \vee \mathbf{1}]$, but note we could not define $\vee$ without the aid of $\mathbf{2}$. Thus, this circularity does not seem to get us anywhere. However, if we needed the species, we could define

$$\mathbf{3} = [\mathbf{2} \vee \mathbf{1}],\ \mathbf{4} = [\mathbf{3} \vee \mathbf{1}], \text{ etc.}$$

The reader should check (and it is not all that pleasant to do so) that <u>arbitrary</u> functions can indeed be defined on these finite species in terms of the constructs already available.

<u>TRANSFINITE CONSTRUCTIONS</u>

The mathematician has the advantage over the "ordinary" mortal (a finite mind) of grasping (some of the properties of) <u>infinite species</u>. Or at least that is his conceit, and the author has no desire to argue against that attitude. Neither does he want to waste the time to glorify this ability but rather wishes to make "visible" the underlying mechanism. What is about to be formulated is hardly original, but the new theory in which it occurs does seem to possess advantages over previous proposals.

The idea is simple. Suppose one is given a species a, then many functions can be defined on this species (assuming some elements are known !). In particular suppose we have a special object 0 (better: 0(a) to show its dependence on a) which will be an element of a yet-to-be specified species $\mathbb{T}$ (better: $\mathbb{T}(a)$). Now we can at one form a function from a into $\mathbb{T}$, namely the constant function $[a \to 0]$. Call this function u for the moment. Why not put $u \in \mathbb{T}$ and assume u and 0 are distinct? This way we can try to form even more functions in $[a \to \mathbb{T}]$, because some new values for functions are now available. And then we want to put those functions in $\mathbb{T}$ and to continue this process in an iterative fashion. There is only one defect with the plan: the functions u already belong to certain species ($u \in [a \to \mathbb{T}]$) and are not available for other jobs. The solution is easy: we send instead of u a proxy called u^+. To be precise, we assume these axioms:

(T1) $\qquad \vdash 0(a) \in \mathbb{T}(a),$

(T2) $\qquad u \in [a \to \mathbb{T}(a)] \vdash u^+ \in \mathbb{T}(a)$.

So far the axioms give positive results about certain elements belonging, but we need more: a principle of (transfinite) induction to assure that these are the only elements (cf.(G5) in the finite case). Here is a possible formulation:

(T3)

$$\frac{\Delta \vdash [0(a)/t][\sigma]\varepsilon[0(a)/t][\tau] \qquad \Delta,\ u\varepsilon[a\to\mathbb{T}(a)],\ \forall x\varepsilon a[[u(x)/t[\sigma]]\varepsilon\ \forall x\varepsilon a[[u(x)/t][\tau]] \vdash [u^+/t][\sigma]\varepsilon[u^+/t][\tau]}{\Delta,\ t\varepsilon\mathbb{T}(a) \vdash \sigma\varepsilon\tau}\ ,$$

where u,t are not free in Δ. We can also take = instead of ε, but this only seems useful for finite a. Nevertheless, let us assume it as (T3'). The lack of extensionality for functions on infinite a may make the use of these axioms somewhat less interesting.

Intuitively (following Tait [18]) the elements of $\mathbb{T}(a)$ can be considered as trees. Thus 0 is the null tree and $[a \to 0]^+$ is the tree of rank $\mathbb{1}$. There are (in general) many trees of rank $\mathbb{2}$. Suppose we take $a = 2$, then any diagram such as :

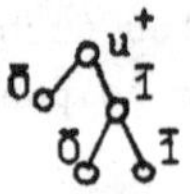

can be considered as determining a function. Here $u = [0 \wedge [0 \wedge 0]]$ and u^+ is our (abstract) tree of rank $\mathbb{2}$. An so on to the higher ranks.

The case $a = 2$ leads to the more beautiful diagrams, but the case $a = \mathbb{1}$ is also of interest (while $a = \mathbb{0}$ is not). Indeed $\mathbb{T}(\mathbb{1})$ can be considered to be the species of <u>integers</u>. Let us write $\mathbb{N} = \mathbb{T}(\mathbb{1})$ and, for the moment, 0 for $0(\mathbb{1})$ and n^+ for $[\mathbb{1} \to n]^+$. Then a moments thought shows us that (T1), (T2), (T3), and (T3') specialize to obvious <u>closure</u> and <u>induction</u> principles for the integers.

Just as with finite sets, we do not have enough functions on the new species unless we assume some additional axioms. In the case of an inductively defined species such as $\mathbb{T}(a)$, the proper method is to supply functions defined by <u>recursion</u>. That is the purpose of the operator $\mathcal{R}$:

(T4) $f = \mathcal{R}\; v\,[a,b,\sigma] \vdash f(0(a)) = b$

(T5) $f = \mathcal{R}_{\,v}\,[a,b,\sigma],\; u \in [a \to \mathbb{T}(a)], v = \forall x \in a[f(u(x))] \vdash f(u^+) = \sigma,$

(T6) $f = \mathcal{R}\; v\,[a,b,\sigma] \vdash f = \forall t \in \mathbb{T}(a)[f(t)].$

This completes the list of fundamental axioms.

As an example of a definition by recursion, we define the important notion of the <u>nodes</u> of a tree. We will define an operation $nd(a) = \forall t \in \mathbb{T}(a)[nd(a)(t)]$, whose values are species. Obviously, what we want are these two theorems:

$$\vdash nd(a)(0(a)) = \mathbb{1}$$

$$u \in [a \to \mathbb{T}(a)] \vdash nd(a)(u^+) = [\mathbb{1} \vee \exists x \in a[nd(a)(u(x))]].$$

The way to obtain them is to define:

<u>DEFINITION</u>

$$nd(a) = \mathcal{R} v[a, \mathbb{1}, [\mathbb{1} \vee \exists x \in a[v(x)]]] .$$

Then the desired theorems will be derived from (T4) and (T5). We can then define such notions as a labeled tree: that is a tree $t \in \mathbf{T}(a)$ together with a function $l = \forall n \in nd(a)(t)[l(n)]$ and so on. (The tree we employ seem less general at first sight than those of TAIT [18], but the idea of a labeled tree is actually more general than TAIT'S notion.)

It may be that the operator $\mathcal{R}$ is defective, because the author can see no way of defining a function (call it pred, say) such that pred $(0) = 0$ and pred $(u^+) = u$. That is, we have not allowed our recursive functions at u^+ to depend not only on the preceding function values, but also on the preceding arguments - the operation pred is a simple example. In the case of $\mathfrak{a}$ finite this seems to be no problem, but for infinite a one sees no easy reduction of the more general kind of recursion to the simpler one. If this is so, we should probably replace $\mathcal{R}$ by $\mathcal{R}'$ with an appropriate axiom such as:

(T5') $$f = \mathcal{R}'u,v[a,b,\sigma],\ u \in [a \to \mathbf{T}(a)],\quad v = \forall x \in a[f(u(x))] \vdash f(u^+) = \sigma .$$

That may look the same as (T5), but note that u is now a bound variable in $\mathcal{R}'u,v[a,b,\sigma]$: that means that u occurs in the same way on both sides of $f(u^+) = \sigma$ as desired.

In case $\mathfrak{a}$ is a finite species, then $\mathbf{T}(\mathfrak{a})$ is denumerable. On the other hand, if $\mathfrak{a}$ is infinite (say $\mathfrak{a} = \mathbf{N} = \mathbf{T}(\mathbf{1})$), then $\mathbf{T}(\mathbf{N})$ is non-denumerable (the ranks of these trees - classically anyway - would be ordinals of the second number class).

We could then go on to iterate $\mathbf{T}$ and form $\mathbf{T}(\mathbf{T}(\mathbf{N}))$, $\mathbf{T}(\mathbf{T}(\mathbf{T}(\mathbf{N})))$, and so on - even into the transfinite. these new species are ever larger and more complicated, but at the moment the author does not really know how to make good use of anything worse than $\mathbf{T}(\mathbf{N})$ - which seems to have been BROUWER'S limit - at least for ordinary analysis. But we agree with TAIT [18] that there appears to be nothing that will stop us at the second step. This, by the way, seems to answer BISHOP [2], who asks whether there is any structure (of a combinatorial rather than function-space nature) beyond the integers. Thus $\mathbf{T}(\mathbf{N})$ is just the index set one needs for the proper definition of BOREL sets, for example, which in [1] where defined by BISHOP only in an intuitive, non-formal way.

INTERPRETING LOGIC

We have already introduced the operators $\forall, \exists, \wedge, \vee, \to$, though the reader may not have yet appreciated just why we used symbols from logic in the way we have. All will now be explained satisfactorily, let us hope. But before we do, we need the notation for the remaining logical operators:

DEFINITIONS

$$\top = \mathbf{1},$$
$$\bot = \mathbf{0},$$
$$\neg[a] = [a \to \bot],$$
$$[a \leftrightarrow b] = [[a \to b] \wedge [b \to a]] .$$

Let us begin with an example of a (valid) logical formula:

$$[\forall x \varepsilon a[[P(x) \to Q(x)]] \to [\neg[\forall x \varepsilon a[Q(x)]] \to \neg[\forall x \varepsilon a[P(x)]]]] .$$

Now as this stands (except maybe for the pedantic bracket conventions and the use of the capital Roman letters P and Q) it can be read both as a formula of ordinary predicate calculus (with variables restricted to a given domain a) and as a term of our theory of constructions. Call the formula $\mathcal{L}$. The reason that $\mathcal{L}$ is intuitionistically (constructively, if you prefer) valid is that there is a specific term τ (involving both of the variables P and Q as well as a) such that the assertion

$$\vdash \tau \varepsilon \mathcal{L}$$

is provable in the theory of constructions.(The blank can be filled in, as we did in several elementary exercises already.) It is just as Professor HEYTING said:
"The proof of the theorem consists in this construction itself, and the steps of the proof are the same as the steps of the mathematical construction".
Of course, we have aided the "intuitively clear mental acts" by our formal rules for operating with constructions. Thus we can ask a machine to check over our proof (as Professor DE BRUIJN wants to do).

We are a little hasty here : the exact term τ required above has not yet been exhibited. It is rather long to write down, and so we shall arrive at it indirectly. Instead of showing why $\mathcal{L}$ is valid (notation: $\vdash \mathcal{L}$), we shall rather establish the validity of this sequent of logical formulas:

$$\forall x \in a[[P(x) \to Q(x)]],\ \neg[\forall x \in a[Q(x)]],\ \forall x \in a[P(x)] \vdash \bot .$$

In general, to establish the validity of a sequent of logical formulas:

$$\mathcal{A}_0, \mathcal{A}_1, \ldots, \mathcal{A}_{n-1} \vdash \mathcal{B}$$

we read them as terms of the theory of constructions and provide <u>both</u> a term τ <u>and</u> a proof in our theory of the assertion:

$$\Delta, t_0 \in \mathcal{A}_0, t_1 \in \mathcal{A}_1, \ldots, t_{n-1} \in \mathcal{A}_{n-1} \vdash \tau \in \mathcal{B}$$

where $t_0,\ldots,t_{n-1}$ are fresh, distinct variables, Δ contains the proper information about the <u>free</u> variables - including the predicate variables - (and more on this later), and where τ may involve all the variables. This means that giving the constructions establishing the $\mathcal{A}_i$ we can always find (in a uniform way by means of τ) a construction which establishes $\mathcal{B}$.

Returning to the example, if we have $t_0 \in \forall x \in a[[P(x) \to Q(x)]]$, $t_1 \in \neg[\forall x \in a[Q(x)]]$ and $t_2 \in \forall x \in a[P(x)]$, then we can certainly find a term $\tau \in \bot$. Indeed, we can let:

$$\tau = t_1(\forall x \in a[t_0(x)(t_2(x))]),$$

and it is a mildly intersting exercise to verify that this **is** correct. (By the way the use of subscripts here is not to be confused with the subscripts $Z_{\mathbf{0}}$ and $Z_{\mathbf{1}}$ for <u>disjoint sums</u> as these subscripts **0 and 1** will be printed in bold-face type.)

As a second example consider the well-known law of contradiction

$$[P \to [\neg[P] \to Q]] .$$

We should try to find τ such that

$$t_0 \in P, t_1 \in \neg[P] \vdash \tau \in Q.$$

This <u>cannot</u> be done unless $\mathbf{0} = \bot$ is really assumed empty, which we are reluctant to do. So we side-step the **issue** by adjoining (as part of the Δ mentioned above) a side condition on the variable Q: namely

$q\varepsilon[\bot\to Q]$. Then $\tau = q(t_1(t_o))$ gives the answer. This should be done for all the propositional variables in general - and for the predicate variables too. Thus if we wanted to establish the validity of :

$$\forall y\varepsilon a[\forall x\varepsilon a[[P(x) \to [\neg[P(x)] \to Q(y)]]]],$$

we would prove :

$$q\varepsilon[\bot\to Q],\ y\varepsilon a,\ x\varepsilon a,\ t_o\varepsilon P(x),\ t_1\varepsilon\neg[P(x)] \vdash \tau\varepsilon Q(y),$$

where $\tau = q(t_1(t_o))(y)$ in this case. Correct? No! we should also have the hypothesis $Q = \forall x\varepsilon a[Q(x)]$ to be able to pass from $q(t_1(t_o))\varepsilon Q$ to $\tau\varepsilon Q(y)$, but that is all quite reasonable.

We can think of Δ as the complete list of the declarations of the types of the variables. In case of predicate, we must indicate their domains and number of arguments, as well as providing for free such $q\varepsilon[\bot\to O]$. We require no more formal statement for Δ for the present.

As a third example we establish the validity of a law of double negation :

$$[\forall x\varepsilon a[[\neg[\neg[P(x)]] \to P(x)]]\to[\neg[\neg[\forall x\varepsilon a[P(x)]]]\to\forall x\varepsilon a[P(x)]]].$$

Thus suppose that $t_o\varepsilon\forall x\varepsilon a[[\neg[\neg[P(x)]]\to P(x)]]$ and that $t_1\varepsilon\neg[\neg[\forall x\varepsilon a[P(x)]]]$. Then as the reader can easily verify - if he has the patience - the construction that belongs to $\forall x\varepsilon a[P(x)]$ is

$$\forall x\varepsilon a[t_o(x)(\forall u\varepsilon\neg[P(x)][t_1(\forall v\varepsilon\forall x\varepsilon a[P(x)][u(v(x))])])] .$$

(The main point of this particular example was to show how complicated and overloaded with brackets the terms can become. This is not to be regarded as a conceptual drawback, however.)

One major point has been left unexplained : our examples were compound formulas, but when we exhibited constructions we broke the implications up into simpler parts. The justification of this procedure lies in the proof of the deduction theorem : if

$$\alpha_o, \alpha_1, \ldots \models \mathcal{L}$$

is valid, then so is

$$\alpha_1, \ldots \models [\alpha_o \to \mathcal{L}] .$$

(Let τ be the term establishing the validity of the first sequent then $\forall t_0 \varepsilon \mathcal{O}_0[\tau]$ - except for a quibble about subscripts - is the term we need to establish the validity of the second sequent.) It may safely be left to the reader to establish modus ponens :

$$\mathcal{O}_0,\ [\mathcal{O}_0 \to \mathcal{O}_1]\ \models \mathcal{O}_1\ ,$$

and since the analogues of all the inference rules (I1)-(I5) for logic are also clearly valid, we have all we need for intuitionistic implication.

Furthermore, the reader, without realizing what he was about (or maybe he did!), has verified in the previous sections all the other axioms of intuitionistic propositional calculus. Thus we can take that as firmly established.

Finishing up the predicate calculus, we note that we have already done the axioms of instantiation. Therefore, it only remains to discuss the rules of generalization :

$$\frac{\mathcal{O}_0,\ \mathcal{O}_1,\ \ldots \models \mathcal{L}}{\mathcal{O}_0,\ \mathcal{O}_1,\ \ldots \models \forall x \varepsilon a[\mathcal{L}]}$$

where x is not free in the $\mathcal{O}_1$, and

$$\frac{\mathcal{O}_0,\ \mathcal{O}_1,\ \ldots \models \mathcal{L}}{\exists x \varepsilon a[\mathcal{O}_0],\ \mathcal{O}_1,\ \ldots \models \mathcal{L}}$$

where this time x is not free in $\mathcal{O}_1, \ldots, \mathcal{L}$. (Let τ be the term establishing the validity of the hypothesis of the first rule. Then $\forall x \varepsilon a[\tau]$ establishes the conclusion. Let σ, on the otherhand, establish the validity of the hypothesis of the second rule. Then $[(t_0)_{\mathbf{0}}/x][(t_0)_{\mathbf{1}}/t_0]\sigma$ is the required term for the conclusion. (Note the bold-face subscripts!))

Though our examples were restricted to monadic formulas, the procedure is quite general; and we can claim that we have given a "mathematical" basis (foundation) for the whole of intuitionistic predicate logic. (By the way, our pedantic notation for binary relationships is P(x)(y), and similarly for more arguments.) It seems quite reasonable to suppose that the proof of LÄUCHLI [14] can be transposed to this theory, and that we can establish the faithfulness of the interpretion (formulas valid on our interpretation are indeed provable in HEYTING'S calculus). The author has not yet had time to work out the details, however, LÄUCHLI did not discuss higher-order logic (though LAWVERE did), and we wish soon to consider a particularly mathematically important theory: higher-order analysis - but without the free choice sequences. Before we do this, however, we must review the progress of our program.

REVIEW

We began by regarding species and constructions as mathematical objects and found that there were some simple axioms governing their properties. It then became slowly apparent that these properties were highly analogous to properties familiar from formal logic. We then turned this analogy into a dogma by insisting that the logical formulas be read (better : interpreted) as (mathematically meaningful) terms of the theory of constructions. This interpretation requires that validity be asserted by the act of giving an explicit construction belonging to the (interpretation of the) formula. Validity is established by giving a proof from the axioms for constructions of the membership assertion.

The next step is to argue that the interpretation is "correct", but so far all we have done is to check the validity of the expected formulas. Thus the situation must be examined in more detail. For one thing : have we verified BROUWER'S "main thesis" ? Which is prior : logic or mathematics? Well, the answer all depends on what one means by logic (what is mathematical seems much clearer). In order to organize the mathematical properties of the constructions into a coherent body of knowledge, we had to set up some rules of deduction (I1) - (I5) and some general axioms such as (E1) - (E2). This represents simply a codification of hypothetical argument (if such and such conditions are fulfilled, then another condition follows). If one can hazard a guess, these principles are so self-evident that BROUWER may never have ever given them a moments' thought. This is the realm of urlogic, without which (in the author's opinion) mathematics (and even coherent thought) is impossible. All of these principles are used naturally in an unconscious fashion. What BROUWER probably meant by "logic" was the elaborate RUSSELL-WHITEHEAD theory of propositional operators, quantifiers and propositional functions and the kind of logic that is meant when RUSSELL says that mathematics is reduced to logic. The author considers that today there is a general agreement that RUSSELL was wrong (or at least over-optimistic). The type-theoretic / set-theoretic foundation for mathematics is not pure logic, because axioms about abstract entities are required - and this falls on the mathematical side of the line between the subjects.

Now what the author feels he has demonstrated here is that - granted urlogic - the combinatorial aspects of (constructive) logic can be given a "mathematical" foundation by the theory of constructions. Of course, the axioms for constructions are not so different from the axioms for logic ((P3) is a rule of universal generalization). Nevertheless a certain reduction has been effected (implication and quantification out of the same operator $\forall$, for example) and a considerable amount of clarity has been gained : one can prove the various propositional formulas from the more elementary principles about constructions.

Professor CURRY hoped for a similar reduction based on his theory of combinators, but the author does not feel that illative combinatory logic (cf. [3] and [4]) has reached a high enough of development to judge it successful.

In particular the author wonders whether CURRY'S long struggle with the many headed monster of the partial function was a serious tactical error. (Note in this connection the remark in footnote 3, p.296 of [4]: "It seems best to proceed with these features (of partial functions) and introduce refinements later in the illative theory".)

One of the author's statements in next to last paragraph requires further discussion : why are properties of constructions "more elementary" than valid propositional formulas? From the point of view of classical two-valued logic this is simply not so. But one must keep in mind that we are investigating propositions here in a constructive way. Thus a proposition does not simply degenerate to one of two truth values but instead is represented by a complex species of possible constructions that conceivably can be used in its validation. From this constructive point of view propositional formulas are not so trivial.

Now what about the interpretations of the logical connectives : are they "correct"? Take implication first. Assuming for simplicity that no hypothesis of declarations are required, what must be done in order to establish $[\mathcal{A} \to \mathcal{B}]$? One must produce a construction together with a proof that this construction transforms every construction that could establish $\mathcal{A}$ into a construction for $\mathcal{B}$.

The construction is an object of the theory while the proof is an elementary argument about the theory. KREISEL [13] calls such proofs 'judgements' and asks for an abstract theory of them. We have not provided this because we did not see why such a theory was needed. The reader may **decide** : have we or have we not carried out the spirit of HEYTING'S interpretation of implication? Of course, this is not new : the KREISEL-GOODMAN theory can be **justified** by a similar discussion. The author only wants to claim that his theory is simple, and as yet that there is no demonstrated need for "abstract" proofs. (But see postscript.)

Conjunction :

To establish $[\mathcal{A} \wedge \mathcal{B}]$ one must produce a pair of constructions the first of which provably justifies $\mathcal{A}$, and the second $\mathcal{B}$.

Disjunction :

To establish $[\mathcal{A} \wedge \mathcal{B}]$ one must produce (another kind of) pair whose first coordinate is either $\bar{0}$ or $\bar{1}$: if $\bar{0}$, then the second coordinate justifies $\mathcal{A}$; if $\bar{1}$, then $\mathcal{B}$.

Truth :

The justification of $\top$ is known because $\dot{0} \in \mathbf{1}$.

Absurdity :

No justification of $\perp (= \mathbf{0})$ is known.

Universal quantification :

To establish $\forall x \varepsilon a [\mathcal{A}]$ one must produce a construction that maps every element of the domain into a justification of the corresponding instance of $\mathcal{A}$.

Existential quantification :

To establish $\exists x \varepsilon a [\mathcal{A}]$ one must produce a pair whose first coordinate is an element of the domain and whose second coordinate provably justifies the corresponding instance of $\mathcal{A}$. This completes our review and our argument for "correctness" (cf. also discussion in MYHILL [16] and in TROELSTRA [20]).

One last topic before we turn to "real" mathematics. KREISEL has often stressed that the reason for having abstract proofs is to make the proof predicate decidable. Otherwise there is no reduction in logical complexity when one says that implication means that _if_ you have a proof of the hypothesis, then you know a proof of the conclusion. In our theory we have replaced _proof_ by _construction_ and _of_ by _membership_. But our theory of membership is a completely 'positive' theory, and we have no way of formulating an assertion to the effect that every object is either a _member_ or _non-member_ of a given species. Likewise, we have no superlarge functions to serve as ($\bar{0}$ - $\bar{1}$ - valued) characteristic functions of species. (TAIT and GOODMAN would allow such functions, but then the door is opened to the murky combinators. So the question is (and it is a quite serious question) : are these deficiencies a real defect of our theory and has the attempted formalization of the basis for intuitionistic logic aborted?

The question of decidability was asked by LACOMBE at the lecture. The author cannot at the moment give a definite answer to this question. The best he can do is to formulate a _conjecture_. You see, from the definition of validity of logical formulas every such assertion can be put in the form :

$$\vdash \sigma\varepsilon\tau \text{ .}$$

(All the variables on the left-hand side of the $\vdash$ can be moved over by (P3) to the right-hand side. Likewise for the side conditions including $p\varepsilon[\bot\to P]$. There only remain the equations of the form $P = \forall x\varepsilon a[P(x)]$; but by (F2) these will disappear, if we substitute $\forall x\varepsilon a[P(x)]$ for P. Of course, the formula is no longer either beautiful or readable, but that is beside the point.) So then, we have the :

FUNDAMENTAL CONJECTURE

There is a (primitive recursive) decision method for the provability in the theory of constructions for assertions of the from $\vdash \sigma\varepsilon\tau$.

Even if the answer is _yes_ to this question, it may not satisfy KREISEL. The decidability is _external_ to the system rather than a condition having an _internal_ formulation. The question may also be related to the "normal-form" problem that de BRUIJN has encountered in his system. It may be that the current proof theoretical work on GÖDEL'S theory $\mathbf{T}$ (by TAIT and HOWARD, among others) sheds light on the problem, because the theories are related. The only thing the author can definitely contribute to the discussion at this moment is that there can be no decision method for assertions of the _hypothetical_ form :

$$\sigma_0 \varepsilon \tau_0, \sigma_1 \varepsilon \tau_1, \ldots, \sigma_{n-1} \varepsilon \tau_{n-1} \vdash \sigma_n \varepsilon \tau_n .$$

We will prove this result even for the fragment based on (F1) - (F2), (P1) - (P3). (This is a stronger not weaker result, because it seems reasonable to suppose that a theorem in the pure theory of functions and products that is proved with the aid of the other notions can be proved without them. But this has not been formally established.) Now if we allowed equations in the hypothesis, the undicidability is immediate : any calculus of conditional equations between arbitrary functions is undecidable in view of the word problem for semigroups. For example, we can easily prove in our system :

$$f \varepsilon [a \to a], g \varepsilon [a \to a], \forall x \varepsilon a [f(g(x))] = \forall x \varepsilon a [g(g(f(x)))], x \varepsilon a \vdash$$
$$f(f(g(x))) = f(g(g(g(f(f(x)))))) .$$

In other words, any deduction from generators and relations written as functional equations can be carried out for the "semigroup" of functions on a domain in our calculus. Now we have not dicussed models for the theory and shall not be able to do so in this paper, but with their aid we can see that, conversely, every equational result proved by pure semigroup methods. Hence, there can be no decision method for the calculus.

We note in passing that the equation in the conclusion - which was written between elements and not between functions in view of the lack of extensionality - could even have been eliminated in favor of membership statements. Thus (and this no doubt can be established with the aid of models) an assertion $\Delta \vdash \sigma = \tau$ is provable if and only if Δ, $h(\sigma) \varepsilon b \vdash h(\tau) \varepsilon b$ is provable, where h and b are new variables. Or even if this is not the case in full generality, enough is true to apply to "semigroup" equations; because for them we need only consider domains with a characteristic function for identity. This simple-minded approach does not, however, eliminate the equations in the hypothesis.

To express deductions with semigroup equations entirely with membership statements, we imagine a function e such that for $x, y \varepsilon a$ the value $e(x)(y)$ is $\top$ in case $x = y$, and is $\bot$ otherwise. We will require nothing special about $\top$ and $\bot$ and could just think of them as free variables - in fact, $\bot$ will not even appear but was just mentioned for definiteness. With this idea about e we recognize several correct statements about it (where we omit some tiresome brackets) :

(i) $\forall x \varepsilon a\ \forall t \varepsilon \top [t] \varepsilon \forall x \varepsilon a [\top \to e(x)(x)]$

(ii) $\forall x \varepsilon a\ \forall y \varepsilon a\ \forall z \varepsilon a\ \forall t \varepsilon \top \forall u \varepsilon e(x)(y) \forall v \varepsilon e(x)(z) [t] \varepsilon \forall x \varepsilon a\ \forall y \varepsilon a\ \forall z \varepsilon a [\top \to [e(x)(y) \to$
$[e(x)(z) \to e(y)(z)]]]$

(iii) $\forall f \varepsilon [a \to a] \forall x \varepsilon a\ \forall y \varepsilon a\ \forall t \varepsilon \top\ \forall u \varepsilon e(x)(y) [t] \varepsilon \forall f \varepsilon [a \to a] \forall x \varepsilon a\ \forall y \varepsilon a [\top \to | e(x)(y) \to$
$e(f(x))(f(y))]]$.

Statements (i) - (iii) express that e is very much like an equality relation on the domain - at least in some formal sense. Next we consider a typical (defining) relation between given (generating) functions f and g for our "semigroup" :

(iv) $\forall x \varepsilon a\ \forall t \varepsilon \top [t] \varepsilon\ \forall x \varepsilon a [\top \to e(f(g(x)))(g(g(f(x))))]$.

We can call (iv) the translation of the 'equation' $fg = ggf$. Now if we let Δ be the sequence (i), (ii), (iii), (iv), $f \varepsilon [a \to a]$, $g \varepsilon [a \to a]$, then it is fairly simple to see that $\Delta \vdash \delta$ can be proved, where δ is the translation of the equation $ffg = ggggff$. Furthermore, if δ is the translation of some other equation, then it is intuitively clear that $\Delta \vdash \delta$ is provable in our calculus if and only if there is a semigroup deduction of the equation from the given $fg = ggf$. Hence, the undecidability result follows with 95% certainty. The status of the conjecture remains open, however. (The undecidability result is not all that interesting, but it is a non-trivial exercise in the theory of constructions that gives some insight into the expressive power of the calculus.)

By the way, in the case where recursion is available in the theory it seems very likely that there is no decision method for assertions $\vdash \sigma \varepsilon \tau$ either. For suppose the term p represents (a standard definition of) a primitive recursive function. Surely there is no way to decide the provability of such assertions as

(*) $$n \varepsilon \mathbb{N} \vdash p(n) = 0 \ .$$

Now let ζ be introduced by recursion so that

$$\vdash \zeta(0) = 0 \ , \text{ and}$$

$$n \varepsilon \mathbb{N} \vdash \zeta(n^+) = \bar{0} \ .$$

It seems reasonable to suppose that

$$\vdash \forall n \in \mathbb{N}[\zeta(p(n))] \in [\mathbb{N} \to \mathbb{N}]$$

is provable if and only if (*) is provable. Hence, there could be no decision method. - This will require some more thought. Maybe there are some sensible restrictions to put on the theory, or maybe one is only interested in special cases of $\vdash_{\sigma\varepsilon\tau}$.

INTERPRETING ANALYSIS

Of course, by analysis we understand higher-order arithmetic, since for foundational purposes we do not need to discuss here the mathematical theory of the real numbers - the reduction of the reals to (sequences of) the integers is assumed known.

We recall that the species of integers $\mathbb{N} = \mathbb{T}(\mathbb{1})$, and that we simplified the notation for 0 and for successors $(n^+ = [\mathbb{1} \to n]^+)$. By recursion we can introduce all the usual primitive recursive functions and prove at least all the basic theorems of primitive recursive arithmetic (eg. GOODSTEIN (7), Chapter V). In particular we can introduce the equality function $E \in [\mathbb{N} \to [\mathbb{N} \to \mathbb{2}]]$ and prove :

$$\vdash E(0)(0) = \bar{0}$$

$$m \in \mathbb{N} \vdash E(0)(m^+) = \bar{1},$$

$$n \in \mathbb{N} \vdash E(n^+)(0) = \bar{1},$$

$$n \in \mathbb{N},\ m \in \mathbb{N} \vdash E(n^+)(m^+) = E(n)(m),$$

$$n \in \mathbb{N},\ m \in \mathbb{N},\ E(n)(m) = \bar{0} \vdash n = m\ .$$

(as will be seen from [7], this is not so easy, but it is elementary.) This allows us to define the predicate of equality between integers :

DEFINITION

$$[n =_{\mathbb{N}} m] = [\top \wedge \bot](E(n)(m))\ .$$

One can then establish the validity of all the usual logical formulas involving equality over the domain $\mathbb{N}$. It takes a little trouble, but let us assume its done.

The next step is to consider the validity of formulas of higher-order arithmetic. What are these formulas? In the first place they contain variables of several sorts or types. We already have at hand the notation for these types :

$$\mathbb{N},\ [\mathbb{N}\to\mathbb{N}],\ [\mathbb{N}\to[\mathbb{N}\to\mathbb{N}]],\ [[\mathbb{N}\to\mathbb{N}]\to\mathbb{N}],$$

and so on. We can imagine what a <u>stratified formula</u> should be (all the types of arguments and values of functions in terms should match). The atomic formulas are numerical (type $\mathbb{N}$) equations, and we may use constants 0, $^{+}$, and anyother well-known functions.

The main effort here is seeing what the formulas are, because <u>validity</u> is already understood (in theory).

We leave to the (poor) reader the verification of :

(A1) $\models \exists n\varepsilon\mathbb{N}\,[[n =_{\mathbb{N}} 0]]$

(A2) $\models \forall n\varepsilon\mathbb{N}\,[\exists m\varepsilon\mathbb{N}[[m =_{\mathbb{N}} n^{+}]]]$

(A3) $\models \forall n\varepsilon\mathbb{N}\,[\,[0 =_{\mathbb{N}} n^{+}]]$

(A4) $\models \forall n\varepsilon\mathbb{N}\,[\forall m\varepsilon\mathbb{N}[[[m^{+} =_{\mathbb{N}} n^{+}]\to[m =_{\mathbb{N}} n]]]]$, .

but they are, after all, rather easy. We shall discuss, however, the <u>induction axiom</u> :

(A5) $\models [P(0)\to[[\forall n\varepsilon\mathbb{N}[[P(n)\to P(n^{+})]]\to\forall n\varepsilon\mathbb{N}[P(n)]]]]$.

For this we must "fill in" the τ of :

$$P = \forall n\varepsilon\mathbb{N}[P(n)],\ p\varepsilon[\bot\to P],\ t\varepsilon P(0),\ u\varepsilon\forall n\varepsilon\mathbb{N}[[P(n)\to P(n^{+})]] \vdash \tau\varepsilon P \quad .$$

(For this particular argument we do not require the $p\varepsilon[\bot\to P]$, but the author wanted to state the problems in full.)

The construction of the construction τ will be given by - recursion, which is hardly surprising. The only trick is to know what the values of the function should be. The function we want has values $[n\wedge\tau(n)]$ (supposing for the moment we already knew our τ), because they can conveniently be chosen by recursion : let

$$\tau' = Rv[\mathbf{1},[0\wedge t],\ [v(\mathring{0})(\bar{0})^{+}\wedge u(v(\mathring{0})(\bar{0}))(v(\mathring{0})(\bar{1}))]] \ .$$

Then $\tau = \forall n \in \mathbf{N}[\tau'(n)(\bar{1})]$ (In the above formula, the reader is reminded that $\mathbf{N} = \mathbf{T}(\mathbf{1})$ and in (T5), $v = [\mathbf{1} \to f(u(\dot{0}))]$. The "trick" of the recursion is to let the next value of the function depend not only on the previous value but also on the previous argument.) The desired result will then be proved by the induction principle (T3). The reasoning is not really circular : we are showing how to reduce a compound form of induction back to a more primitive kind. Nevertheless, we do not do away with <u>all</u> assumptions. (Likewise, in set theory we still need an axiom of infinity to have a <u>set</u> of integers.)

Next we have the <u>axiom of choice</u> :

(A6) $\quad \forall x \in a[\exists y \in b[P(x)(y)]] \models \exists f \in [a \to b][\forall x \in a[P(x)(f(x))]]$,

where we are using not only a free binary predicate variable P, but also free "type" variables a and b as well.

Let $\pi = Pf\in[a \to b][\forall x \in a[P(x)(f(x))]]$. Then for τ take $\tau = \pi(q)(\psi)$, where we assume $t \in \forall x \in a[\exists y \in b[P(x)(y)]]$ and let $\varphi = \forall x \in a[t(x)_0]$ and $\psi = \forall x \in a[t(x)_1]$. It is that easy, because the interpretation of the existential quantifiers is so constructive.

Rather more complicated is the axiom of dependent choices, a principle very important for analysis but curiously overlooked until recently (cf.eg. the end of MYHILL [17]) :

(A7) $\forall x \in a[[P(x) \to \exists y \in a[[P(y) \wedge R(x)(y)]]]] \models$

$\forall x \in a[[[P(x) \wedge Q(x)] \to \exists f \in [\mathbf{N} \to a][[[P(f(0)) \wedge Q(f(0))] \wedge \forall n \in \mathbf{N}[R(f(n))(f(n^+))]]]]]$.

With our bracketing conventions, this is an axiom that is harder to write than to understand. In words : if a <u>finite</u> chain of relationships among elements of a having property P can be indefinitely extended, then (assuming we have a constructive verification of the hypothesis) we can find an <u>infinite</u> chain of elements with successive terms of the chain related and the initial element specified. (The reader should not overlook the generalizations of this principle to trees other than those in $\mathbf{T}(\mathbf{1})$, but this is neither the time nor place to discuss them.) The desired sequence of elements is found by recursion, but one must be careful on which species the recursion is done.

To verify (A7), note that the hypothesis is equivalent (in a sense to be made precise in a moment) to the "formula" :

$$\forall u\varepsilon\, \exists x\varepsilon a[P(x)][\, \exists v\varepsilon\, \exists x\varepsilon a[P(x)]\, [R(u_0)(v_0)]\ .$$

The sense of equivalence is simply this : given a construction belonging to the hypothesis of (A7), we can find a construction belonging to the above - and conversely. (This is the meaning of $\leftrightarrow$, too.) So let t be a construction belonging to the above. As in the argument for (A6) consider the two functions $\varphi = \forall u\varepsilon\, \exists x\varepsilon a[P(x)]\, [t(u)_0]$ and $\psi = \forall u\varepsilon\, \exists x\varepsilon a[P(x)]\, [t(u)_1]$. From the assumption on t we note that we can prove in the theory the conclusion

$$\psi\varepsilon \forall u\varepsilon\, \exists x\varepsilon a[P(x)]\, [R(u_0)(\varphi(u)_0)]\ ,$$

which is just a more explicit version of our main hypothesis if we also remember

$$\varphi\varepsilon[\, \exists x\varepsilon a[P(x)] \rightarrow \exists x\varepsilon a[P(x)]]\ .$$

Next suppose that $Z\varepsilon a$ and $p\varepsilon P(Z)$ and $q\varepsilon Q(Z)$. We then iterate φ by recursion obtaining $\tilde{\varphi}$ so that

$$\tilde{\varphi}\varepsilon[\mathbb{N} \rightarrow \exists x\varepsilon a[P(x)]]$$
$$\tilde{\varphi}(0)_0 = Z,\ \tilde{\varphi}(0)_1 = p, \text{ and}$$
$$\tilde{\varphi}(n^+) = \varphi(\tilde{\varphi}(n)),$$

for $n\varepsilon\mathbb{N}$. (We are speaking informally, but this can all be done in the system.) Next we let $f = \forall n\varepsilon\mathbb{N}[\tilde{\varphi}(n)_0]$, and we find

$$p\varepsilon P(f(0)),\ q\varepsilon Q(f(0)), \text{ and}$$
$$\psi(\tilde{\varphi}(n))\varepsilon R(f(n))(f(n^+))$$

for $n\varepsilon\mathbb{N}$. By using all manner of pairing functions all these facts can be put together to obtain a term which can finally be shown to belong to the conclusion of (A7). (A note to the reader who tries this : remember $\tilde{\varphi}$ contains Z and p as free variables and that you will have to apply functional abstraction to them.)

Very much analysis can already be carried out on the basis of (A1) - (A7) (using intuitionistic logic!) and we have BISHOP [1] as evidence. However, such topics as BOREL sets and continuous functions on BAIRE space (eg. functions of type $[[\mathbb{N}\to\mathbb{N}]\to\mathbb{N}]$) bring up definitions by recursion on the second number class or better : on $\mathbf{T}(\mathbb{N})$. We shall only discuss one topic here : the definition of the predicate K on $[[\mathbb{N}\to\mathbb{N}]\to\mathbb{N}]$ that determines the continuous functions and which is obtained by recursion.

Before giving the recursion on $\mathbf{T}(\mathbb{N})$, it is convenient to introduce by an ordinary recursion on $\mathbb{N}$ that operator * such that

$$m\varepsilon\mathbb{N},\ f\varepsilon[\mathbb{N}\to\mathbb{N}] \vdash [m * f](0) = m\ ,$$

and

$$m\varepsilon\mathbb{N},\ f\varepsilon[\mathbb{N}\to\mathbb{N}],\ n\varepsilon\mathbb{N} \vdash [m * f](n^+) = f(n)\ .$$

This is simple and we need not give the explicit definition for $*$. One can think of $*$ as a kind of translation operator on BAIRE space.

To define K we define an auxiliary operator $\hat{K}$ by recursion on $\mathbf{T}(\mathbb{N})$ such that :

$$\vdash \hat{K}(0(\mathbb{N})) = \forall k\varepsilon[[\mathbb{N}\to\mathbb{N}]\to\mathbb{N}][\exists n\varepsilon\mathbb{N}[\forall f\varepsilon[\mathbb{N}\to\mathbb{N}][[k(f) =_{\mathbb{N}} n]]]],$$

and

$$u\varepsilon[\mathbb{N}\to\mathbf{T}(\mathbb{N})] \vdash \hat{K}(u^+) = \forall k\varepsilon[[\mathbb{N}\to\mathbb{N}]\to\mathbb{N}]$$
$$[\forall m\varepsilon\mathbb{N}[\hat{K}(u(m))(\forall f\varepsilon[\mathbb{N}\to\mathbb{N}][k([m * f])])]].$$

Then K can be defined by the equation :

$$K = \forall k\varepsilon[[\mathbb{N}\to\mathbb{N}]\to\mathbb{N}][\exists t\varepsilon\mathbf{T}(\mathbb{N})[\hat{K}(t)(k)]].$$

The intention of the definition is that a construction in K(k) when $k\varepsilon[[\mathbb{N}\to\mathbb{N}]\to\mathbb{N}]$ gives the direct evidence of why k is continuous. The motivation for the definition is based on the well-known inductive analysis of continuous functions on $[\mathbb{N}\to\mathbb{N}]$. Unfortunately, we do not have the time to discuss the notion further here but can only mention the axioms that can be validated, namely those of <u>closure</u> and <u>induction</u>:

(A8) (i) $\vdash \forall n\varepsilon\mathbb{N}[K([[\mathbb{N}\to\mathbb{N}]\to n])],$

(ii) $\vdash \forall k\varepsilon[\mathbb{N}\to[[\mathbb{N}\to\mathbb{N}]\to\mathbb{N}]][\forall m\varepsilon\mathbb{N}[K(k(m))]\to$
$$K(\forall f\varepsilon[\mathbb{N}\to\mathbb{N}][k(f(0))(\forall n\varepsilon\mathbb{N}[f(n^+)])])]],$$

$$
\begin{aligned}
&\text{(iii)}\ \forall n \in \mathbb{N}[P([[\mathbb{N}\to\mathbb{N}] \to n])],\\
&\quad \forall k \in [\mathbb{N}\to[[\mathbb{N}\to\mathbb{N}] \to \mathbb{N}]][[\forall m \in \mathbb{N}[P(k(m))] \to\\
&\qquad\qquad P(\forall f \in [\mathbb{N}\to\mathbb{N}][k(f(0))(\forall n \in \mathbb{N}[f(n^{+})])])]]\\
&\quad \vdash \forall k \in [[\mathbb{N}\to\mathbb{N}] \to \mathbb{N}][[K(k) \to P(k)]].
\end{aligned}
$$

Clearly, if we do not soon introduce some abbreviations, our formulas will be quite impossible to read. The worst part of the above (A8) is the clumsy restrictions of the variables.

Suppose we let $\mathbb{B} = [\mathbb{N}\to\mathbb{N}]$ and $\mathbb{F} = [\mathbb{B}\to\mathbb{N}]$. Further let us define $\bar{f} = \forall n \in \mathbb{N}[f(n^{+})]$. Then for example (A8) (ii) reads :

$$\vdash \forall k \in [\mathbb{N}\to\mathbb{F}][\forall m \in \mathbb{N}[K(k(m))] \to K(\forall f \in \mathbb{B}[k(f(0))(\bar{f})])],$$

where we have also left out some brackets. It could have been shortened even further if we had given a special name to the transformation :

$$\forall k \in [\mathbb{N}\to\mathbb{F}][\forall f \in \mathbb{B}[k(f(0))(\bar{f})]] \in [[\mathbb{N}\to\mathbb{F}] \to \mathbb{F}].$$

Another approach to practical readibility would be to have conventions that certain variables were to range over certain species. It is hard to stick to these conventions when our alphabet is so finite, however.

This completes our brief survey of the foundations of analysis based on the theory of constructions. What we have given should have been enough, though, to convince the reader that our theory is a sufficiently strong and fertile one.

CONCLUSION

We have tried to present here with adequate motivation a theory of constructions and to show how it is in harmony with BROUWER'S program - at least as the program has been explained by HEYTING. We consider the attempt rather successful, but much remains to be done. For example, we have not discussed quantification over species (better : subspecies of a given species). This can be done in a convenient way within the framework of the present theory, though it is necessary to adjoin new primitive notions. Such considerations bring up problems of consistency, and we have not had time here to investigate the many interesting models that can be (non-constructively!) fashioned for the theory. Especially interesting is a model (of which the author is 85% sure that it can be defined) that has the property that all functions are continuous. This thesis, which is certainly related to BROUWER'S view - except we are not making use of choice sequences - ought to have rather interesting consequences. But there are other thesis possible too : we might want to assume that all basic species are countable - in the sense that they can all be mapped one-one into $\mathbb{N}$. Clearly that thesis also would have different but far reaching consequences. And then there is CHURCH'S Thesis and KRIPKE'S Schema, and these should all be investigated further. What we have accomplished here is the providing of a good context within which to compare these assumptions.

In another direction, we find a host of proof-theoretical problems. One must transpose LÄUCHLI'S proof to this context as well as the results of GOODMAN'S Thesis. A point to think about is whether the author has made the transfinite part of the theory too strong. Are there theorems of first-order arithmetic that can be validated using constructions based on $\mathbb{T}(\mathbb{N})$ but not without it? And what about $\mathbb{T}(\mathbb{T}(\mathbb{N}))$? And what is the strength of the theory with only finite species? A different question : does the constructive proof of GÖDEL'S Incompleteness Theorem suggest any reflection principles that could be added to the theory preserving its constructive character? Would this be a way in which an "abstract" theory of proofs might become interesting again? There seem to be quite a number of things to think about in this area, and the theory of constructions - in this form or another - gives us a way of making the questions and answers precise.

POSTSCRIPT

After further discussions with KREISEL and GÖDEL it has become clear that the attempt to eliminate "proofs" (as abstract objects) and to concentrate on the "pure" constructions is not successful : the decidability problems definitely show that the desired reduction of logical complexity has not been obtained. Therefore, the theory must be revised. (For an exact formulation of a relevant "adequacy condition", as KREISEL calls it, see Problem 10 of his [13].) The author is still unable to formulate any "abstract" theory of proofs that would seem convenient, but he has a suggestion that might be sufficient for the purpose of an adequate theory of constructions. Namely, we replace the elementary assertions $\Delta \vdash \delta$ by assertions $\varepsilon : \Delta \vdash \delta$ where ε is a term denoting a construction which measures the stage at which $\Delta \vdash \delta$ can be proved. Many people have considered stages of evidence, and it seems as though the constructions can easily be used to index these stages and to formalize the idea. For one thing proofs (and ordinals) can be related to trees (as BROUWER did himself) and as we noted above, the constructions can also be thought of as trees. The idea will require some development, and the author did not want to publish this paper until he was more certain that the approach is reasonable. But maybe the details we have outlined here can be of some inspiration to others.

(1) BISHOP, E.A.
Foundations of constructive analysis - New-York, 1967.

(2) ——
Mathematics as a numerical language, Buffalo Conference, 1968, to appear.

(3) CURRY, H.B. and FEYS, R.
Combinatory logic, vol. I - Amsterdam 1958.

(4) CURRY, H.B.
Combinatory logic, in Contemporary Philosophy, a survey. (R.Klibansky, ed.) Florence, 1968, pp. 295 - 307.

(5) GOODMAN, N.D.
Intuitionistic arithmetic as a theory of constructions. Thesis, Stanford University, 1968.

(6) ——
A theory of constructions equivalent to arithmetic. Buffalo Conference, 1968, to appear.

(7) GOODSTEIN, R.L.
Recursive number theory - Amsterdam, 1957.

(8) HEYTING, A.K.
Intuitionism in mathematics, in Philosophy in the mid-century, a survey (R.Klibansky,ed.) Florence, 1958, pp. 101 - 115.

(9) KLEENE, S.C. and VESLEY, R.E.
Foundations of intuitionistic mathematics - Amsterdam, 1965.

(10) KREISEL, G.
Foundations of intuitionistic logic, in Logic, Methodology, and the Philosophy of Science (E. Nagel, P. Suppes, A. Tarski, eds.) Stanford, 1962, pp. 198 - 210.

(11) ——
Mathematical logic, in lectures on modern mathematics, vol. 3 (T.L. Saaty, ed.) New-York, 1965, pp. 95 - 195.

(12) ——

Functions, ordinals, species, in Logic, Methodology and Philosophy of Science III (B. van Rootselaar and J.F. Staal, eds.) Amsterdam 1968, pp. 145 - 159.

(13) ——

Church's thesis : a kind of reducibility axiom for constructive mathematics, Buffalo Conference, 1968, to appear.

(14) LÄUCHLI, H.

An abstract notion of realizability for which intuitionistic predicate calculus is complete, Buffalo Conference, 1968, to appear.

(15) LAWVERE, F.W.

Category-valued higher-order logic, UCLA Set Theory Institute, 1967,to appear.

(16) MYHILL, J.

Notes towards a formalization of intuitionistic analysis, Logique et Analyse, vol. 35, 1967, pp. 280 - 297.

(17) ——

Formal systems of intuitionistic analysis I, in Logic Methodology and Philosophy of Science III (B. van Rootselaar, J.F. Staal, eds.) Amsterdam, 1968, pp. 161 - 178.

(18) TAIT, W.W.

Constructive reasoning, in Logic, Methodology, and Philosophy of Science III (B. van Rootselaar, J.F. Staal, eds.) Amsterdam, 1968, pp. 185 - 199.

(19) TROELSTRA, A.S.

The theory of choice sequences, in Logic, Methodology and Philosophy of Science III (B. van Rootselaar, J.F. Staal, eds.) Amsterdam, 1968, pp. 201 - 223.

(20) ——

Principles of intuitionism, Lecture Notes in Mathematics vol. 95 Springer Verlag 1969.

Paramodulation and Set of Support*

Lawrence Wos • George Robinson

INTRODUCTION

The applications of the set-of-support strategy in fields in which automatic theorem proving plays an important role continue to increase. Question-answering systems such as that of Green [3] and information-retrieval systems such as that of Darlington [2] rely heavily on automatic theorem proving. Since one of the principal problems in theorem proving is generation of a very large number of "irrelevant" inferences, this problem is important for any system based on a theorem-proving subprogram. The set-of-support strategy was formulated to impede the generation of irrelevant inferences and thus restrict the number of inferences to be examined during the search for a refutation. It is desirable however for the restricted system to retain refutation completeness.

Two inference systems will be considered in this paper: the system Π employing the inference rules paramodulation, resolution, and factoring, and the system Σ employing only resolution and factoring. The system Π is applicable to first-order theories with equality, while Σ is efficiently applied only to ordinary first-order theories. ΠT and ΣT are identical to Π and Σ respectively, except that only T-supported inferences are allowed. The system ΣT (Σ with T as set of support) has been shown to be refutation complete when S is finite and S-T is satisfiable. (The finiteness consideration for S is not necessary for refutation-completeness, but is necessary to show that a certain procedure is a refutation procedure.)

*Work performed under the auspices of the U. S. Atomic Energy Commission.

A number of papers have been published modifying the original resolution principle. Such modifications include hyper-resolution [8], semantic resolution [9], and resolution with merging [1]. Due no doubt in part to the success with which the set-of-support strategy has met, a common point of interest has been the question of refutation completeness of the system combining set of support with the particular modification of resolution then under consideration. For such systems the conclusion usually is, briefly speaking, that set of support is complete. More formally, for such modified systems Ω, ΩT is frequently refutation complete.

In this paper the definition of *set of support* is extended to inference systems based on paramodulation, thus extending the scope of automatic theorem-proving from ordinary first-order theories to first-order theories with equality, the former being the subject of earlier theorem-proving papers. The question which naturally comes to mind is answered in the main result of the paper: set of support is refutation complete for functionally reflexive first-order theories with equality.

DEFINITIONS AND NOTATION

In this paper A, B, C, D, E, and F will be clauses; R will be the equality predicate; S, T, and U will be (not necessarily finite) sets of (not necessarily ground) clauses; f will be a function symbol; k, l, and m will be literals; s, t, and u will be terms; $x_1, x_2, \ldots$ will be individual variables; σ and τ will be substitutions of (not necessarily ground, i.e., not necessarily variable-free) terms for variables; and Ω will be an inference system.

Definition (Paramodulation): Let A and B be clauses (with no variables in common) such that a literal Rst (or Rts) occurs in A and a term u occurs in (a particular position in) B. Further assume that s and u have a most general common instance $s' = s\sigma = u\tau$ where σ and τ are the most general substitutions such that $s\sigma = u\tau$. Where $\hat{B}$ is obtained by replacing by $t\sigma$ the occurrence of $u\tau$ in the position in $B\tau$ corresponding to the particular position of the occurrence of u in B, infer (from any variants A* and B* of A and B respectively) the clause $C = \hat{B} \cup (A - \{Rst\})\sigma$ (or $C = \hat{B} \cup (A - \{Rts\})\sigma$). C is called a *paramodulant* of A and B (and also of B and A) and is said to be *inferred by paramodulation from* A *on* Rst (or Rts) *into* B *on (the occurrence in the particular position in* B *of)* u. The literal Rst (or Rts) is called the *literal of paramodulation* [6].

From a given pair A and B of clauses one can usually infer by paramodulation a number of clauses. Which clause is inferred depends first on the direction of paramodulation (A into B or B into A), then on the equality literal of paramodulation, then on the choice (first or second) of argument within that literal, then on the term and its occurrence within the other clause.

Notation: $S\mathcal{P}T$ will be $\{C \mid C$ can be inferred by paramodulation from A into B or from B into A where $A \varepsilon S$ and $B \varepsilon T\}$. "$\{A\}\mathcal{P}\{B\}$" will be abbreviated "$A\mathcal{P}B$"; "$S\mathcal{P}\{B\}$" by "$S\mathcal{P}B$", etc.

For example, where R is the equality predicate and, intuitively, f is product and g inverse, if A is Rf(yg(y))e and B is Rf(xx)e, then $A\mathcal{P}B$ = {{Rf(ef(yg(y)))e}, {Rf(f(yg(y))e)e}, {Rf(xx)f(yg(y))}, {Rf(eg(f(xx)))e},{Rf(f(xx)g(e))e}, {Rf(yg(y))f(xx)}}. The first three elements of $A\mathcal{P}B$ are obtained by paramodulation from A into B, and the last three from B into A.

Since the terms *resolution* and *resolvent* vary somewhat in usage throughout the literature, we give the following:

Definition: For any literal l, $|l|$ is that atom such that either $l = |l|$ or $l = -|l|$.

Definition (Resolution): If A and B are clauses (with no variables in common) with literals k and l respectively such that k and l are opposite in sign (i.e., exactly one of them is an atom) but $|k|$ and $|l|$ have a most general common instance m, and if σ and τ are most general substitutions with $m = |k|\sigma = |l|\tau$, then infer from (any variants A* and B* of) A and B the clause $C = (A - \{k\})\sigma \cup (B - \{l\})\tau$. C is called a *resolvent* of A and B and is inferred by *resolution* [4][7].

Definition (Factoring): If A is a clause with literals k and l such that k and l have a most general common instance m, and if σ is a most general substitution with $k\sigma = l\sigma = m$, then infer the clause $A' = (A - \{k\})\sigma$ from A. A' is called an *immediate factor* of A. The *factors* of A are given by: A is a factor of A, and an immediate factor of a factor of A is a factor of A.

As with paramodulation, resolution yields a number of possible inferences from a given pair A and B of clauses. SRT will be $\{C \mid C$ is a resolvent of $A \in S$ and $B \in T\}$, etc. FS will be $\{C \mid C$ is a factor of $A \in S\}$, etc.

Definition: If R is the equality predicate, a set S of clauses is *functionally reflexive* if $Rxx \in S$ and if, for each n-ary function f occurring in S, $Rf(x_1,\ldots,x_n)f(x_1,\ldots,x_n) \in S$.

The theories of interest are the functionally reflexive first-order theories with equality. The scope of interest would be *all* first-order theories with equality if it were not for the fact that refutation completeness for Π without functional reflexivity is an open question. If and when the corresponding theorem is proved, it would be desirable to extend the results which follow to all first-order theories with equality.

Since the inference systems which play the main role throughout are Π (the inference system consisting of paramodulation, resolution, and factoring) and ΠT (identical to Π except that only clauses with T as set of support are allowed), it becomes necessary to extend the definition of set of support [10] to include inferences made through paramodulation.

Definition: Given a set S with subset T, a clause C has *T-support* (with respect to S) if $C \in T$, or if C is a factor of a clause with T-support, or if C is a paramodulant or resolvent of clauses A and B where B has T-support and either A has T-support or is a factor of a clause in S-T. T is called a *set of support* for C.

The definition could be further extended to any inference system Ω by replacing paramodulation, resolution, and factoring by "some rule of Ω".

That the definition of set of support given above is no more than an extension of the definition given in [10] can be seen by examining the alternate definition for "C having T-support" given below. In the definition below, T_S^i is extended from that given in [10].

<u>Definition</u>: S^o is the set of clauses B such that B is in S or there is a clause C in S with B a factor of C. For $i > 0$, S^i is the set of clauses A such that $A \varepsilon S^{i-1}$, or there exist clauses $C \varepsilon S^{i-1}$ and $D \varepsilon S^{i-1}$ such that A is a paramodulant or a resolvent of C and D or A is a factor of a paramodulant or resolvent of C and D.

<u>Definition</u>: For $T \subseteq S$, T_S^o is the set of clauses A such that $A \varepsilon T$ or such that A is a factor of some clause B with $B \varepsilon T$. For $i > 0$, T_S^i is the set of clauses A such that $A \varepsilon T_S^{i-1}$, or there exist clauses $C \varepsilon T_S^{i-1}$ and $D \varepsilon S^o \cup T_S^{i-1}$ such that A is a paramodulant or a resolvent of C and D or A is a factor of a paramodulant or a resolvent of C and D.

Since the factors of a clause A include A itself as a trivial factor, S^o consists of the factors of the clauses of S. When S contains only ground clauses, it is obvious that $S^o = S$. Normally, however, S contains nonground clauses, and in many such cases $S^o - S$ is not empty. (From the fact that A is a factor of itself it follows that some of the definitions given above can be appropriately shortened.)

<u>Definition</u>: The *S-level* of a clause A (relative to Ω) is the smallest i such that $A \varepsilon S^i$. The *T_S-level*[1] of A is the smallest i with $A \varepsilon T_S^i$.

Since, for all clauses A, A is a factor of itself, T_S^i for $i > 0$ can be obtained from T_S^{i-1} by adjoining to T_S^{i-1} all clauses E which are factors of some clause D where D is in turn inferrable by paramodulation or resolution from some pair B and F with B in T_S^{i-1} and F in the $S^o \cup T_S^{i-1}$.

---

[1]That which is now termed T_S-level was formerly termed T-level in some of our earlier papers.

Definition: Given a set S of clauses, a subset T of S, and a clause A deducible from S, A is said to have *T-support* if, for some $i \geq 0$, $A \in T_S^i$. T is said to be a *set of support for A,* and A is said to be *supported by T.*

Definition: A *T-supported deduction* $D_1, D_2, \ldots, D_n$ (relative to S and Ω) is a deduction in Ω in which every D_i has T-support in Ω or is a factor of a clause in S-T. If such a deduction exists we write $S \vdash_{\Omega T} D_n$.

Definition: A set S of clauses is *R-satisfiable* if it has an R-model, i.e., a model in which the predicate R is mapped to an equality relation.

Definition: A *refutation* of S is a deduction from S of the empty clause, $\square$.

Definition: An inference system Ω (or ΩT) is *R-refutation complete* if for R-unsatisfiable S, $S \vdash_{\Omega} \square$ (or $S \vdash_{\Omega T} \square$).

Definition: If $T \subseteq S$ and $S \vdash_{\Omega} C$, then C has *T-heritage* (relative to S and Ω) if in Ω there is no deduction of C from S-T (i.e., $S\text{-}T \nvdash_{\Omega} C$).

The concept of T-heritage bears an interesting relation to the concept of T-support as evidenced by Lemmas 5 and 6. T-heritage is a concept which has in the past been confused with T-support; this point and related ones will be clarified in the next section. That the concept of T-heritage is distinct from the concept of T-support can be seen from the following example:

> Let A = {-P,-Q,R}, B = {P,Q}, C = {P,-Q}, S = {A,B,C}, T = {C}. F = {Q,-Q,R} is a (tautologous) resolvent of A and B, and D = {P,-Q,R} is a resolvent of F and C. D has T-heritage, but D is not in T_S^i for any i and, therefore does not have T-support.

MISCONCEPTIONS AND NON-EQUIVALENT DEFINITIONS OF SET OF SUPPORT

It is incorrect, as can be seen from the example given below, to restate casually the heart of the definition of set of support as follows: If C is inferrable by paramodulation or resolution from A and B, and if at least one of A and B has T-support and both are deducible from S, then C has T-support.

The example under consideration is that given at the end of the previous section. The clause D does not have T-support even though one of its parents, C, does. As has been said, D has T-heritage, and there exists by Lemmas 5 and 6 a subclause E of D such that E has T-support. The only element of (CRB)RA will do for E (as can be seen by examining the proof of Lemma 1).

We give an additional example to show that the casual rendering of the set of support definition given above can lead to an error when both paramodulation and resolution are involved as rules of inference.

Let A = {Rab,-Qc}, B = {Pa,Qc}, C = {Pa,-Qc}, S = {A,B,C}, T = {C}. D, the only element of (APB)RC, is {Pa,Pb,-Qc}. Although D has T-heritage, D does not have T-support even though one of its parents does.

The proof of Lemma 3 gives the clause E = {Pb,-Qc}, which is a clause whose existence is demanded by Lemmas 5 and 6. E has T-support and is a subclause of D. E is the only element of (CRB)PA.

The question of T-support status for some given clause D is in general only semidecidable even if S is finite. Although one can have a decidable test for D being an element of a given T_S^j (the union of T_S^0, $T_S^1, \ldots, T_S^j$ is finite for each j), all that can be said in general is that, if D has T-support, then this fact can be ascertained eventually since D will be in some T_S^i. If D does not have T-support, the situation is analogous to attempting to prove that a given non-theorem is in fact a non-theorem.

The question of T-heritage for a given clause is also in general only semidecidable. (Putting the set of support question another way, one normally cannot show that D is not in T_S^i for all i.)

For us if a clause is in some T_S^i it has T-support regardless of whether or not it is deducible from S-T.

Slagle [9] demands[2] that, in order for a deduction to have T-support, no resolution occurs between members of S-T (ignoring factoring for this discussion). Thus all of his T-supported deductions are for us also T-supported, but not conversely as can be seen from the following example:

$S-T = \{A,B,C,E\}$, $T = \{F\}$, $A = \{P,R\}$, $B = \{P,-R\}$, $C = \{Q,R\}$, $E = \{Q,-R\}$, $F = \{-P,Q\}$. $D_1 = \{P,R\}$. $D_2 = \{P,-R\}$. $D_3 = \{-P,Q\}$. $D_4 = \{Q,R\}$, a resolvent of D_1 and D_3. $D_5 = \{Q,-R\}$, a resolvent of D_3 and D_2. $D_6 = \{Q\}$, a resolvent of D_4 and D_5.

The deduction D_1 through D_6 has T-support for us, but not for Slagle since he does not allow the resolution of D_4 and D_5, both of which are in

[2] He also assumes S-T satisfiable, which is irrelevant to what follows and is mainly done because of his intended application; we wish not to make this assumption because of the generality gained and because of other applications by other authors such as Green [3] concerning question-answering systems.

S-T. This resolution is allowable for us because D_4 and D_5 have T-support since they are elements of T_S^1. Although Slagle does not define set of support for clauses but instead only for deductions, he would in effect exclude {Q} from having T-support while {Q} would have T-support for us. He would in effect generate each T_S^i, but before retaining it remove from it all elements already in S-T.

The reason for such attention to this difference in definition is twofold. First of all, one should note that his refutation completeness theorem is strictly stronger than that given in [10]. Secondly, since Slagle's definition allows fewer deductions, (smaller T_S^i), it might seem best to prove in this paper the stronger refutation completeness theorem as his approach might be more efficient.[3] The proof of Lemma 5, however, breaks down immediately since, even with F in S-T one cannot conclude that the elements of CRF or CPF have T-support when C does since some or all of such elements may also be in S-T.

Even with the obvious possible modification Lemma 5 is false for Slagle. For a counter-example, let S-T consist of the three clauses {P,R}, {Q,-R}, {-R,S}, and T consist of the clause {-Q,S}. D = {P,S} is a clause satisfying the hypothesis of Lemma 5 and, therefore, for us must have a subclause with set of support. D itself for us has T-support, but no subclause

---

[3] Slagle's definition of set of support corresponds, at least on the unit level, to that which has been programmed in PG1 through PG5. Besides the stronger completeness theorem, he has shown (unpublished) that an instance C' of a clause C in S-T can be discarded without losing refutation completeness even when C' has T-support. For unit clauses this result has been used for a number of years in the programs PG1 through PG5.

of D exists either in S-T or obtainable with a T-supported deduction in the sense of Slagle.

The question of whether or not Lemma 6 holds with Slagle's definition of T-support is at the present an open question. The example just given does not serve as a counter-example since the clause D of the example does not have T-heritage.

Lemmas 5 and 6 may give real insight into the question, intuitively speaking, of why set of support is refutation complete for ΠT (in the presence of functional reflexivity) and ΣT.

LEMMAS, COROLLARIES, AND THEOREMS

Lemmas 1 to 6 are reordering lemmas with 1 to 4 being local and 5 to 6 global. All six are proved on the ground level here, although analogous lemmas are probably provable on the non-ground level if factoring is appropriately utilized. Lemmas 7 and 8 are used to obtain a non-ground refutation from a given ground-clause refutation and are so-called "capturing lemmas" for factoring and resolution. The obvious analog to Lemma 8, but with resolution replaced by paramodulation, is frequently not true. (For a counter-example, let A = Rab = A', B = Qx, B' = Qg(a), C' = Qg(b); the only factor respectively of A and B are A and B themselves, $A\mathcal{P}B$ consists of {Qa} and {Qb}. There is, therefore, no C in $E\mathcal{P}F$ with C' as an instance, see Lemma 8.) The lack of a paramodulation "capturing lemma" analogous to 8 has been *the* source of difficulty in proving refutation completeness of paramodulation-based inference systems when functional reflexivity was not assumed [6].

For the proofs of Lemmas 1 through 4, note that $\mathcal{P}$ and $\mathcal{R}$ are symmetric: $S\mathcal{P}T = T\mathcal{P}S$ and $S\mathcal{R}T = T\mathcal{R}S$ for all sets S and T. Note also that the relation of "ancestry" is one between occurrences of literals rather than between literals themselves. When the proof calls for the paramodulation of a pair of clauses on a pair of literals, it is intended that the choice both of direction of paramodulation and of term occurrence is dictated by the history of the case under study unless specifically otherwise stated.

Lemma 1. If D is a clause in $(A\mathcal{R}B)\mathcal{R}C$ then there exists a subclause E of D with $E \in (C\mathcal{R}B)\mathcal{R}A \cup (C\mathcal{R}A)\mathcal{R}B \cup (C\mathcal{R}B)\mathcal{R}(C\mathcal{R}A)$.

Proof. Let D be in (ARB)RC. Then there exists a clause F ε ARB such that D is a resolvent of F and C. F and C must, therefore, contain complementary literals, say q in F and -q in C. Similarly, there exist literals p in B and -p in A such that F is inferrable by resolution from B and A on p and -p. D is inferred from F and C on q and -q. Since q ε F, q ε B or q ε A (or both). If q is in B, and if q = -p or q is not in A, then, where G is the resolvent of C and B on -q and q, let E be the resolvent of G and A on p and -p. E ε (CRB)RA and is a subclause of D. If q is in both B and A, and if q ≠ -p, then, where G is as above and H is the resolvent of C and A on -q and q, let E be the resolvent of G and H on p and -p. E ε (CRB)R(CRA) and is a subclause of D. The remaining case yields a subclause E of D with E ε (CRA)RB. The proof is complete.

<u>Lemma 2.</u> If D ε (APB)PC, then there exists a subclause E of D with E ε (CPB)PA ∪ (CPA)PB ∪ ((CPA)PB)PC ∪ ((CPB)PA)PC.

Proof. Let D ε (APB)PC. Then there exists F ε APB with D a paramodulant of C and F.

Case 1. D is inferred by paramodulation from F into C. Let $\tilde{r}_2$ ε F be the (equality) literal of paramodulation. Since F ε APB, depending on whether paramodulation was from A into B or from B into A, one of A and B contains the (equality) literal, say r_1, of paramodulation and the other contains the literal, say p_1, containing the term occurrence of paramodulation. Since $\tilde{r}_2$ ε F, there exists a literal r which is an ancestor of $\tilde{r}_2$ in A or B (or both). r ≠ $\tilde{r}_2$ precisely when r is that literal p_1 which is involved in inferring F in the discussion above.

Case 1a. There exists an ancestor r_2 of $\tilde{r}_2$ such that r_2 ε B and $\tilde{r}_2 = r_2$. Let G be inferred by paramodulation on $\tilde{r}_2$ ε B into p_2 ε C, where p_2 contains the term occurrence in the paramodulation of C and F to get D. The literals of G are, with one possible exception, elements of D. The possible exception is the literal (r_1 or p_1) from B. The only literal of B which may not be in G is $\tilde{r}_2$.

If F was obtained by paramodulation from B into A, then r_1 was in B, and r_1 is not equal to $\tilde{r}_2$ (since $r_1 \varepsilon$ B is deleted in the inferring of F, so could not be an ancestor of $\tilde{r}_2$ in B). So r_1 would be in G. Paramodulate G into A on r_1 and p_1 to infer H. The only literal which may be in H and not in D is $\tilde{r}_2$. If this is not the case, let E be H. If it is the case, let E be inferred by paramodulating H into C on $\tilde{r}_2$ and p_2. Thus, if F was inferred by paramodulating B into A, there exists an E satisfying the theorem with E in the union of (C$\mathcal{P}$B)$\mathcal{P}$A and ((C$\mathcal{P}$B)$\mathcal{P}$A)$\mathcal{P}$C.

Now consider the case in which F was inferred by paramodulating A into B. If $p_1 \notin$ G, p_1 must equal $\tilde{r}_2$ since $\tilde{r}_2$ is the only literal which may be in B and not in G. But then, from the hypothesis of 1a., p_1 is unchanged by paramodulating A into B. So r_1 must be of the form Rtt for some term t. Let $\tilde{p}_2 \varepsilon$ G be the descendant of $p_2 \varepsilon$ C. If $\tilde{p}_2$ contains t as a term, paramodulate A into G on r_1 and $\tilde{p}_2$. Let H_1 be the resulting inference. If H_1 is a subclause of D, let E be H_1. If not, then $\tilde{r}_2$ is in H_1 and not in D. Then paramodulate H_1 into C on $\tilde{r}_2$ and p_2. Thus, if $\tilde{p}_2$ contains t as a term, the desired E is in (C$\mathcal{P}$B)$\mathcal{P}$A $\cup$ ((C$\mathcal{P}$B)$\mathcal{P}$A)$\mathcal{P}$C.

If $\tilde{p}_2$ does not contain t as a term, then p_2 must since t is a term of $\tilde{r}_2$. Then let G_1 be the paramodulant of A into C on r_1 and p_2. Since r_1 = Rtt in the case under discussion the descendant of p_2 in G_1 is p_2. Let H_2 be the paramodulant of B into G_1 on $\tilde{r}_2$ and p_2. The only literal which can be in H_2 and not in D is $\tilde{r}_2$. If not, let E be H_2. If such is the case, let E be the paramodulant of H_2 into C on $\tilde{r}_2$ and p_2. E ε (C$\mathcal{P}$A)$\mathcal{P}$B $\cup$ ((C$\mathcal{P}$A)$\mathcal{P}$B)$\mathcal{P}$C.

The last subcase to consider is where $p_1 \in G$. If $\tilde{r}_2 \in A$ and $\tilde{r}_2 \neq r_1$, let H_3 be the paramodulant of A into G on r_1 and p_1. Then let E be the paramodulant of H_3 into C on $\tilde{r}_2$ and p_2. If $\tilde{r}_2 = r_1$ or $\tilde{r}_2 \notin A$, again let H_3 be the paramodulant of A into G on r_1 and p_1. Let $\tilde{p}_1 \in H_3$ be the descendant of p_1. The only literal of H_3 in this case which may not be in D is $\tilde{p}_1$. If this is not the case, let E be H_3. If it is the case, then $\tilde{p}_1 = \tilde{r}_2$. Then let E be the paramodulant of H_3 into C on $\tilde{r}_2$ and p_2. $E \in ((CPB)PA)PC \cup (CPB)PA$.

Case 1b. There exists an ancestor r_2 of $\tilde{r}_2$ such that $r_2 \in A$ and $r_2 = \tilde{r}_2$. In this case there exists a subclause E of D with $E \in (C\mathcal{P}A)\mathcal{P}B \cup ((C\mathcal{P}A)\mathcal{P}B)\mathcal{P}C \cup (C\mathcal{P}B)\mathcal{P}A \cup ((C\mathcal{P}B)\mathcal{P}A)\mathcal{P}C$. The argument parallels that of 1a.

Case 1c. No ancestor of $\tilde{r}_2$ is equal to $\tilde{r}_2$, but there exists an ancestor r_2 of $\tilde{r}_2$ with r_2 in B. It follows that $r_2 = p_1$, and that either $\tilde{r}_2 \notin A$ or $\tilde{r}_2 = r_1$. There exists, therefore, an argument u_1 of r_2 such that u_1 is replaced by $\tilde{u}$ in inferring F. Since the literal of paramodulation of F and C is $\tilde{r}_2$, either $\tilde{u}$ or u_2, the other argument of r_2, may be the argument being "matched" with a term in $p_2 \in C$. u_2 is unchanged in passing from B to F in all cases since all clauses herein are ground clauses. If u_2 is the argument for match, then let G be the paramodulant of B into C with literal of paramodulation r_2 in B, using u_2 as the match argument. $p_2 \in C$ becomes $\tilde{p}_2 \in G$. Let E be the paramodulant of A into G on r_1 and $\tilde{p}_2$. E is a subclause of D and is in $(C\mathcal{P}B)\mathcal{P}A$. On the other hand, if $\tilde{u}$ is the match argument for F and C, then an argument of r_1 can be successfully matched with the term in p_2. Let H be the resulting inference from A and C, and let p_3 be the transform of p_2. Let E be the paramodulant of B into H on r_2 and p_3, using $u_1 \in r_2$ as the argument for match, where $r_2 = Ru_1u_2$ or $r_2 = Ru_2u_1$. E is a subclause of D and is in $(C\mathcal{P}A)\mathcal{P}B$.

Case 1d. No ancestor of $\tilde{r}_2$ equals $\tilde{r}_2$, but there exists an ancestor r_2 of $\tilde{r}_2$ with $r_2 \in A$. By paralleling the argument of 1c, we obtain a subclause E of D with $E \in (C\mathcal{P}A)\mathcal{P}B \cup (C\mathcal{P}B)\mathcal{P}A$.

Case 2. D is inferred by paramodulation from C into F. Thus there exists a literal r_2 in C of paramodulation and a literal $\tilde{p}_2$ in F containing the term occurrence. Let r_1 and p_1 be the literals for inferring F from A and B. There exists an ancestor of $\tilde{p}_2$ in A or in B or in both.

If there exists an ancestor p_2 of $\tilde{p}_2$ such that $p_2 \varepsilon$ B and $p_2 = \tilde{p}_2$, we can argue as in 1a. If F was inferred by paramodulating from B into A, then the desired E exists in $(CPB)PA \cup ((CPB)PA)PC$. If F was inferred by paramodulating A into B, let G be the paramodulant of C into B on r_2 and $\tilde{p}_2$. If $p_1 \not\varepsilon$ G, then r_1 = Rtt as in 1a. If u is the term of paramodulation in B used for inferring F, and if u is not involved in the inference of G, the desired E is in $(CPB)PA \cup ((CPB)PA)PC$. If u is involved in the inference of G, E is in $(CPA)PB \cup ((CPA)PB)PC$. However, in this last case if G was inferred by paramodulating on a proper subterm of t in u, one must paramodulate from C into A rather than from A into C as in 1a. Finally, if $p_1 \varepsilon$ G, $E \varepsilon (CPB)PA \cup ((CPB)PA)PC$.

If there exists an ancestor p_2 of $\tilde{p}_2$ with $p_2 \varepsilon$ A and $p_2 = \tilde{p}_2$, we argue as in 1b. $E \varepsilon (CPA)PB \cup ((CPA)PB)PC \cup (CPB)PA \cup ((CPB)PA)PC$.

If no ancestor of $\tilde{p}_2$ equals $\tilde{p}_2$, but there exists an ancestor p_2 of $\tilde{p}_2$ with $p_2 \varepsilon$ B, as in 1c there is an E which is a subclause of D and is in $(CPB)PA \cup (CPA)PB$. The argument parallels the subcases of 1c. One may, however, be forced to paramodulate from C into A rather than from A into C as was required at the end of the first subcase of case 2.

If no ancestor of $\tilde{p}_2$ equals $\tilde{p}_2$, but there is an ancestor p_2 of $\tilde{p}_2$ in A, then the desired E is in $(CPA)PB \cup (CPB)PA$. The proof is complete.

<u>Lemma 3</u>. If $D \varepsilon (APB)RC$, then there exists a subclause E of D with $E \varepsilon (CRB)PA \cup (CPA)RB \cup ((CRA)PB)RC \cup (CRA)PB \cup (CPB)RA \cup ((CRB)PA)RC$.

Proof. Let D be a clause in (APB)RC. Then there exists an F ε APB such that D is in FRC. Thus there exist literals $\tilde{q}$ in F and $-\tilde{q}$ in C with $D = (F - \{\tilde{q}\}) \cup (C - \{-\tilde{q}\})$. As in the proof of Lemma 2, we can conclude that there exist literals q_1 in A or B as ancestor of $\tilde{q}$, r_1 and p_1 (one in A, the other in B) with F a paramodulant of A and B on r_1 and p_1 and with D ε CRF.

Case 3a. There exists an ancestor q of $\tilde{q}$ in B such that $q = \tilde{q}$. Let G be the resolvent of C and B on $-\tilde{q}$ and $\tilde{q}$.

If F was obtained by paramodulating B into A, let H be the paramodulant of G into A on r_1 and p_1. If H is a subclause of D, let E be H. If not, then the only literal in H and not in D is $\tilde{q}$. Then let E be the resolvent of H and C on $\tilde{q}$ and $-\tilde{q}$.

If F was obtained by paramodulating A into B, and if $p_1 \notin G$, then r_1 = Rtt for some term t and $p_1 = \tilde{q}$. Let G_1 be the paramodulant of A into C on r_1 and $-\tilde{q}$. Let H_1 be the resolvent of G_1 and B on $-\tilde{q}$ and $\tilde{q}$. If H_1 is a subclause of D, let E be H_1. If not, then $\tilde{q}$ ε A and $\tilde{q} \neq r_1$. In this case $\tilde{q}$ is the only literal in H_1 and not in D. Then let F_1 be the resolvent of C and A on $-\tilde{q}$ and $\tilde{q}$. Let F_2 be the paramodulant of F_1 into B on r_1 and p_1. Let E be the resolvent of F_2 and C on $\tilde{q}$ and $-\tilde{q}$.

If F was obtained by paramodulating A into B and if p_1 ε G, let H_2 be the paramodulant of A into G on r_1 and p_1. If H_2 is a sublcause of D, let E be H_2. If not, then the only literal in H_2 and not in D is $\tilde{q}$. Then let E be the resolvent of H_2 and C on $\tilde{q}$ and $-\tilde{q}$.

In case 3a, we can find a subclause E of D with

E ε (CRB)PA $\cup$ ((CRB)PA)RC $\cup$ (CPA)RB $\cup$ ((CRA)PB)RC.

Case 3b. There exists an ancestor q of $\tilde{q}$ with $q \in A$ and $q = \tilde{q}$. Then, by arguing as in 3a, there exists a subclause E of D with $E \in (C\mathcal{R}A)\mathcal{P}B \cup ((C\mathcal{R}A)\mathcal{P}B)\mathcal{R}C \cup (C\mathcal{P}B)\mathcal{R}A \cup ((C\mathcal{R}B)\mathcal{P}A)\mathcal{R}C$.

Case 3c. No ancestor of $\tilde{q}$ is equal to $\tilde{q}$, but there exists an ancestor q of $\tilde{q}$ with $q \in B$. Then r_1 is in A and is of the form Rst for terms s and t. In obtaining F, q becomes $\tilde{q}$ by replacing the appropriate occurrence of s by t (or by replacing the appropriate occurrence of t by s). Let G_2 be the paramodulant of A into C on r_1 and $-\tilde{q}$. Let E be the resolvent of G_2 and B on -q and q, which is possible since -q is the descendant in G_2 of $-\tilde{q}$ in C. E is a subclause of D since, in the case under discussion, no ancestor of $\tilde{q}$ equals $\tilde{q}$. $E \in (C\mathcal{P}A)\mathcal{R}B$.

Case 3d. If no ancestor of $\tilde{q}$ equals $\tilde{q}$, but an ancestor q of $\tilde{q}$ is in A, $E \in (C\mathcal{P}B)\mathcal{R}A$. The proof is complete.

Lemma 4. If $D \in (A\mathcal{R}B)\mathcal{P}C$, then there exists a subclause E of D with $E \in (C\mathcal{P}B)\mathcal{R}A \cup (C\mathcal{P}B)\mathcal{R}(C\mathcal{P}A) \cup (C\mathcal{P}A)\mathcal{R}B$.

Proof. Let q and -q be respectively in A and B as required for $F \in A\mathcal{R}B$ with $D \in F\mathcal{P}C$, for arbitrary D. If D was obtained by paramodulating C into F, C contains the (equality) literal, say r_1, of paramodulation and F contains the literal, say p_1, containing the term of paramodulation for inferring D. If an ancestor of p_1 is in B, let G be the paramodulant of C into B on r_1 and p_1. If $p_1 \notin A$ or $p_1 = q$, let E be the resolvent of G and A on -q and q. If $p_1 \in A$ and $p_1 \neq q$, let H be the paramodulant of C into A on r_1 and p_1. Let E be the resolvent of G and H on -q and q. Thus, in the case under discussion, a subclause E of D can be found in $(C\mathcal{P}B)\mathcal{R}A \cup (C\mathcal{P}B)\mathcal{R}(C\mathcal{P}A)$. If B contains no ancestor of p_1, then A must. In that case a subclause E of D exists in $(C\mathcal{P}A)\mathcal{R}B$.

If D was obtained by paramodulating F into C, by paralleling the argument just given but with the roles of p_1 and r_1 interchanged one can show the existence of a subclause E of D with $E \in (CPB)RA \cup (CPB)R(CPA) \cup (CPA)RB$. The proof is complete.

<u>Lemma 5</u>. Let S and $T \subseteq S$ be given and let U be the smallest set containing S-T such that U is closed both under paramodulation and resolution. (Factoring is irrelevant on the ground level.) If $F \in U$, and C has T-support, and if $D \in CPF \cup CRF$, then there exists a clause H such that H is a subclause of D and, more importantly, H has T-support.

Proof. Let $(S\text{-}T)^{\circ} = S\text{-}T$ (since ground clauses have no non-trivial factors), and for $j \geq 0$ let $(S\text{-}T)^{j+1} = (S\text{-}T)^{j} \cup APB \cup ARB$ for all clauses A and B in $(S\text{-}T)^{j}$.

Then $U = \bigcup_{j} (S\text{-}T)^{j}$. Let F be a clause in U, C a clause with T-support and D a clause in the union of CPF and CRF. The proof proceeds by induction on the (S-T)-level of F, where the (S-T)-level n of F is (as given earlier) the smallest n such that $F \in (S\text{-}T)^{n}$. If the (S-T)-level of F is 0, then $F \in S\text{-}T$ and D by definition has T-support since C has and F is a paramodulant or a resolvent of C and a clause in S-T. Assume by induction that the lemma is true for clauses G with (S-T)-level j with $0 \leq j \leq n$, and let F be of (S-T)-level n+1. Then there exist clauses A and B in $(S\text{-}T)^{n}$ with $F \in APB \cup ARB$. D, therefore, is in the union of (ARB)RC, (APB)PC, (APB)RC and (ARB)PC. Depending on which of the just given four sets contains D, one of Lemmas 1 through 4 applies to yield a subclause E of D. In addition one knows that E is itself contained in some union of sets dependent on C, B, and A, and on some combination of paramodulation and resolution. We shall give the argument for the case in which $E \in ((CRA)PB)RC$ and show that a subclause H of E and hence a subclause of D, exists and has T-support. The remaining cases can be proved by an argument similar to that which follows but less involved.

Since in the case under discussion E is assumed in ((CRA)PB)RC, there exist clauses G_1 and G_2 with $G_1 \in CRA$, $G_2 \in G_1PB$, and $E \in G_2RC$. Since C has T-support and $A \in (S-T)^n$, by induction there exists the clause E_1 which is a subclause of G_1 and has T-support. E_1 is either itself a subclause of G_2 or E_1 contains the literal relevant to the paramodulation of G_1 and B. In the first case, let $E_2 = E_1$. In the second, apply the induction hypothesis to E_1 and B to show that there exists an E_2 which is a subclause of G_2 and which has T-support. Thus in either case we have a T-supported subclause E_2 of G_2. Either E_2 is a subclause itself of E or contains the literal for resolution with C corresponding to that by which E was inferred. Since E_2 has T-support and since every resolvent of E_2 and C has T-support (for they both do), we have an E_3 which has T-support and is a subclause of E which is a subclause of D. $E_3 = E_2$ or is in E_2RC. $H = E_3$ is the desired subclause of D having T-support.

Lemmas 5 and 6 are proved with ΠT as the underlying inference system. The obvious modification of those proofs will give corresponding lemmas for ΣT.

<u>Lemma 6</u>. Given S and $T \subseteq S$ and a clause D with T-heritage (relative to S and Π) then there exists a subclause E of D such that E has T-support.

Proof. Let D be a clause with T-heritage, not deducible with paramodulation and resolution from S-T but deducible from S with those same inference rules. The proof proceeds by induction on the S-level of D. If D has S-level 0, then $D \in T$ since D has T-heritage. So D itself has T-support. Assume by induction that all clauses with S-level less than or equal to n having T-heritage possess a subclause having T-support, and let D have S-level n+1. Then there exist clauses C and F of S-level less than or equal to n such that $D \in CPF \cup CRF$. If neither C nor F have T-heritage then both are deducible in Π from S-T, and D, therefore, is deducible in Π from S-T. But D has T-heritage in Π, which would be a contradiction, so one of C and F say C has T-heritage. By the induction hypothesis, C has a subclause C_1 having T-support.

If F has T-heritage, then by the induction hypothesis there exists a subclause F_1 of F having T-support. Now C and F with paramodulation or resolution yielded D. Hence C and F each contain a literal relevant to this paramodulation or resolution. If either C_1 or F_1 (subclauses respectively of C and F) lack that particular literal, then the clause lacking the literal is a subclause of D and has T-support from the above. If both C_1 and F_1 have the literals in question then paramodulation or resolution of C_1 and F_1 on that literal pair yields a subclause of D. This subclause has T-support since C_1 and F_1 both do.

Now consider the case where F does not have T-heritage and is, therefore, deducible in Π from S-T. Paramodulation or resolution can be applied to C_1 and F on the literal pair used to infer D, unless C_1 is a subclause of D. In the latter case we are finished. In the former we infer from F and C_1 the clause G which, since C_1 is a subclause of C, is a subclause of D. We apply Lemma 5 to F, C_1, and G to obtain a subclause C_2 of G. C_2 is, therefore, a subclause of D, and by Lemma 5 it has T-support.

It has already been shown by example (see the end of the section on definitions and notation) that the concept of T-heritage is distinct from that of T-support. It follows that, given a clause with T-heritage, Lemmas 5 and 6 may yield at best a proper subclause having T-support. In the example just cited the clause with T-heritage was {P,-QR}, and the subclause provided by Lemmas 5 or 6 is {-Q,R}.

Remark. By examining the proofs of Lemmas 5 and 6, one can prove the corresponding lemmas with Π replaced by Σ. The correspondent of 6 states that, if D has T-heritage relative to S and Σ, there is a subclause E of D which has T-support relative to S and Σ. A similar statement is the correspondent to Lemma 5. The heart of the matter is Lemma 1, which guarantees the existence of a clause (ground) inferrable by resolution with some appropriate re-ordering when presented with a clause D in (ARB)RC.

Definition: Ω is *R-sound* if, whenever $S\vdash_{\Omega}C$, C holds in all R-models of S in which C is defined. An *R-model* of S is a model of S in which R (the equality predicate) is mapped to an equality relation.

Corollary 1. If S is an unsatisfiable set of ground clauses with $T \subseteq S$ such that S-T is satisfiable, then $S\vdash_{\Sigma T}\Box$ (set of support is ground refutation complete in Σ).

Proof. Let S be an unsatisfiable set of ground clauses with $T \subseteq S$ such that S-T is satisfiable. Since resolution is refutation complete, $S\vdash_{\Sigma}\Box$, the empty clause is deducible from S. Since resolution is sound and S-T satisfiable, the empty clause has T-heritage (relative to S and Σ). By the remark above, there exists a subclause of $\Box$ having T-support relative to S and Σ. (Contradictory units could have been the focus of attention instead.) Thus set of support is ground refutation complete in Σ.

Corollary 2. If S is an R-unsatisfiable set of ground clauses and if $T \subseteq S$ is such that S-T is R-satisfiable and if S contains all clauses of the form Rtt for t in the Herbrand universe of S, then $S\vdash_{\Pi T}\Box$; set of support is ground refutation complete in Π.

Proof. Let S and T $\subseteq$ S satisfy the hypothesis of the corollary. Since Π has been shown to be refutation complete for such an S [11] [5], and since paramodulation and resolution are both R-sound and S-T is R-satisfiable, the empty clause has T-heritage (relative to Π and S). Apply Lemma 6

Paramodulation, though R-sound (i.e., sound for first order theories with equality), is of course not sound for ordinary first order theories. Qb is a paramodulant of Qa and Rab but not a logical consequence of Qa and Rab. In Corollary 2 it is not sufficient to require S-T to be satisfiable rather than R-satisfiable as can be seen from the following example.

$$S\text{-}T = \Big\{\{Qa\}, \{-Qb\}, \{Rab\}\Big\}, \text{ and } T = \{Pa\}.$$

S is R-unsatisfiable. S-T is satisfiable but R-unsatisfiable. Obviously there is no T-supported refutation relative to S and Π.

<u>Lemma 7</u>. If A' is an instance of A, then there exists a factor B of A having the same number of literals as A' and having A' as an instance.

<u>Lemma 8</u>. If A' and B' are instances respectively of A and B and if C' is an element of A'$\mathcal{R}$B', then there exists a clause C ε E$\mathcal{R}$F having C' as an instance, where E is a factor of A and F is a factor of B.

Lemmas 7 and 8 are true for all instances and not just for ground instances.

For the theorem which follows, the proof is one of obtaining a deduction based on a set S of clauses from a ground clause deduction based on a set of instances of S.

Occurrences of terms in two literals are said to be in the *same position* if each is the i_n-th argument of the i_{n-1}-st argument of...of the i_1-st argument of its literal.

Lemma 9. If A' and B' are ground instances of clauses A and B respectively, and if C' is a paramodulant of A' into B', and if B has a term in the position corresponding to the term of paramodulation in B', then there exists a clause C that is a paramodulant of some factor of A and some factor of B and such that C' is an instance of C [12].

Theorem 1. If S is a functionally-reflexive R-unsatisfiable set of (not necessarily ground) clauses and if $T \subseteq S$ with S-T R-satisfiable, then $S \vdash_{\Pi T} \square$; set of support is R-refutation complete (in ΠT) for functionally reflexive sets.

Proof. Let H be the Herbrand universe of S, and let S' be the full instantiation of S over H. Since S is R-unsatisfiable, S' is R-unsatisfiable. S' contains all clauses of the form Rtt for all terms t in H, where R is the equality predicate. Let the full instantiation of S-T over H be (S-T)'. (S-T)' is R-satisfiable since S-T is. T', the full instantiation of T over H, is such that S'-T' is R-satisfiable since $S'-T' \subseteq (S-T)'$. By Corollary 2 there exists a T'-supported refutation $D_1', D_2', \ldots, D_n'$. Using the refutation $D_1', D_2', \ldots, D_n'$, the following procedure yields a T-supported refutation $D_1, D_2, \ldots, D_h$ of S itself within Π.

For each clause D_i' of the ground refutation, the procedure yields a finite sequence U_i of clauses such that the last clause in U_i has D_i' as an instance and also has precisely the same number of literals as D_i'. The juxtaposition of $U_1, U_2, \ldots, U_n$ will be a T-supported refutation $D_1, D_2, \ldots, D_h$ in Π of S. In many cases h will be greater than n. This results from two causes: the need for factoring or the need for extra steps due to the lack of a capturing lemma of the desired type for paramodulation.

For each D_i' whose justification is that D_i' is in T', there exists an A in T with D_i' as an instance. If A and D_i' have the same number of literals, let $U_i = A$. If not, let U_i be A,B as provided by Lemma 7.

For each D_i' whose justification is that D_i' is in $S'-T'$, there exists an A in S-T with D_i' as an instance. If A and D_i' have the same number of literals, let $U_i = A$. If not, let U_i be A,B as provided by Lemma 7.

For each D_i' whose justification is that D_i' is a resolvent of D_j' and D_k', consider the corresponding U_j and U_k. Let A be the last element of U_j and B be the last element of U_k. Since by construction A, B, D_j', and D_k' satisfy the hypothesis of Lemma 8, there exists a C in ERF having D_i' as an instance, where E and F are respectively factors of A and B. A, however, has the same number of literals as D_j' by construction. One can deduce, therefore, that E = A. Similarly, it follows that F = B. If C and D_i' have the same number of literals, let $U_i = C$. If not, apply Lemma 7 to obtain the clause D such that D has the same number of literals as D_i' and has D_i' as an instance. Then let U_i be C,D.

For each D_i' whose justification is that D_i' is a paramodulant of D_j' and D_k', let the paramodulation be, without loss of generality, from D_j' into D_k'. Let A and B be respectively the last elements of the corresponding U_j and U_k. Let u' in D_k' be the term occurrence of paramodulation relevant to the inference of D_i'. If B contains a term in the corresponding position to that of u' in D_k', then by Lemma 9 there exist factors E and F of A and B respectively such that some C in EPF has D_i' as an instance. Since by construction, D_j' and A have the same number of literals and similarly D_k' and B have the same number of literals, E = A and F = B. If C and D_i' have the same number of literals, let U_i be C. If not, let U_i be C,D, where D is obtained by applying Lemma 7 to D_i' and C. When B does not have a term in the position corresponding to that of u' in D_k', the property of functional reflexivity comes into play.

In the case now under discussion D_i' is inferred by paramodulation from D_j' into D_k', A and B are respectively the last elements of U_j and U_k, and B does not have a term u in the position (within the corresponding literal) corresponding to u' (the term of paramodulation) in D_k'. Normally A and B will not have a paramodulant C having D_i' as an instance. Depending on whether or not D_k' has T'-support, there are two alternatives for obtaining a pair of clauses one of whose paramodulants has D_i' as an instance.

Consider the case in which D_k' has T'-support. We show that a clause B_r can be inferred by paramodulation from B and a set of functional reflexivity axioms. B_r will have D_k' as an instance and will also have a term u in the position corresponding to u'. Since D_k' is an instance of B, there is a substitution σ such that $B\sigma = D_k'$. Since B lacks a term in the position corresponding to u', there exists a variable x in B and a non-empty set of functions $f_1, f_2, \ldots, f_p$ in D_k' such that σ contains $f_1(\ldots f_2(\ldots f_p(\ldots u' \ldots)\ldots))/x$, i.e., x is replaced in passing from B to D_k' by instantiation by $f_1(\ldots f_2(\ldots f_p(\ldots u' \ldots)\ldots))$. The vector giving the position of x in B is an initial segment of that for u', i.e., the vectors agree on the first q coordinates where q is the number of coordinates giving the position of x in B. Let $G_1, G_2, \ldots, G_p$ be the functional reflexivity axioms corresponding to $f_1, f_2, \ldots, f_p$. Let B_1 be the paramodulant of G_1 into B on x. In general for $1 \leq m \leq p-1$, let B_{m+1} be the paramodulant of G_{m+1} into B_m on the variable occurring as the t-th argument of f_m where t is the (q+m+1)-st coordinate of the position vector for u'. By construction B_1 and, therefore, all B_m,

$1 \leq m \leq p$, have the same number of literals as D_k'. Let $B_r = B_p$. A, B_p, D_j' and D_k' satisfy the hypothesis of Lemma 9. So there exists a clause C in EPF having D_i' as an instance, where E and F are respectively factors of A and B_p. As earlier in the proof one can conclude that E = A and F = B_p since B_p and D_k' have the same number of literals. If C and D_i' have the same number of literals let U_i be $G_1,G_2,\ldots,G_p,B_1,B_2,\ldots,B_p,C$. If not, apply Lemma 7 to C and D_i' to obtain a clause D having D_i' as an instance and having the same number of literals as D_i'. Then let U_i be $G_1,\ldots,C,D$.

The other alternative occurs when D_k' does not have T'-support. But then, since the ground deduction has T'-support, D_j' has T'-support. Now the procedure just given would, of course, still yield a clause D_p with most of the desired properties. But, because of the consideration of T-support desired for $D_1,D_2,\ldots,D_h$, B_p will not do. We shall show instead, therefore, that there exists a clause A_p which can be inferred from A, the last element of U_j, and the G_r, the functional reflexivity axioms of the previous paragraph such that paramodulation from A_p into B yields a clause having a subclause of D_i' as an instance. First we re-number the G_r. For $1 \leq m \leq p$ with r+m = p+1, let H_m be G_r. Since the inference of D_i' was by paramodulation from D_j' into D_k', D_j' contains an equality literal Rs't'. Since the term of paramodulation in D_k' relevant to this inference was u', and since D_j' and D_k' are ground clauses, s' or t' = u'. Without loss of generality say that s' = u'. Since by construction D_j' is an instance of A (the last element of U_j), A contains a corresponding equality literal Rst. A_p will be identical to a subclause of A except that Rst will be replaced by $Rf_1(f_2(\ldots f_p(s)\ldots))f_1(f_2(\ldots f_p(t)\ldots))$, and the last p coordinates of the position vector of s will be identical to the last p for u'. Let $A_o = A$, and A_t for $1 \leqq t \leqq p$ be the paramodulant of A_{t-1} into H_t. Each paramodulation is into the second argument of the corresponding H_t and into the r-th subargument therein where r is the (q+p+1-t)-th coordinate of the position vector of u'. The literals of paramodulation are respectively, Rst, $Rf_p(s)f_p(t),\ldots,Rf_2(f_3(\ldots f_p(s)\ldots))f_2(f_3(\ldots f_p(t)\ldots))$. As was seen earlier there exists the variable x occurring in the position in B whose first q position coordinates are identical to the first q of u' in D'_k. Let B_1 be obtained by paramodulation of A_p into B on $Rf_1(f_2(\ldots f_p(s)\ldots))f_1(f_2(\ldots f_p(t)\ldots))$ into the above mentioned occurrence of x. Let x_1 be the term occurrence resulting from the replacement of that occurrence of x. Let B_2 be the paramodulant of A on Rst into t in x_1 in B_1, where t in its literal in B_1 and u' in its literal in D'_k are in the same position, and let x_2 be the corresponding resulting term occurrence--x_2 replaces x_1. Let B_3 be obtained by paramodulation of A on Rst into s in x_2 in B_2, where s in its literal in B_2 and u' in its literal in D'_k are in the same position. B_3 has D'_i as an instance. If B_3 and D'_i have the same number of literals, let U_i be $H_1,H_2,\ldots,H_p,A_1,A_2,\ldots,A_p,B_1,B_2,B_3$. If not, apply Lemma 7 to B_3 and D'_i to obtain B_4 such that B_4 has D'_i as an instance and B_4 and D'_i have the same number of literals. Then let U_i be $H_1,\ldots,B_3,B_4$.

The need for application of paramodulation to A and B_1 and then to A and B_2 is that certain literals needed to capture D'_i as an instance of A might be lost in passing from A to A_p and then to B_1.

Now generate in order $U_1,U_2,\ldots,U_n$ from $D'_1,D'_2,\ldots,D'_n$ by applying the procedure given above. The desired (possibly) non-ground deduction $D_1,D_2,\ldots,D_h$ is obtained by juxtaposing $U_1,U_2,\ldots,U_n$. Since D'_n is the empty clause and, by the construction, the last element of U_n has D'_n as an instance, $D_1,D_2,\ldots,D_h$ is a refutation. (That $D_1,D_2,\ldots,D_h$ is a deduction including justifications follows from the construction.)

The argument that $D_1, D_2, \ldots, D_h$ is a T-supported refutation proceeds by induction. We show that for each U_i, all elements of U_i are either factors of clauses in S-T or have T-support, and in addition we show that, for all i, if D'_i has T'-support, then the last element of U_i has T-support. D'_1 is in S'-T' or in T'. So by construction the elements of U_1 are either factors of clauses in S-T or factors of clauses in T. If D'_1 has T'-support then by construction the last element of U_1 is a factor of a clause in T. Now assume by induction that our statement holds for U_t with $1 \leqq t \leqq r$, and consider U_{r+1}.

If D'_{r+1} is justified by being in S'-T', then U_{r+1} has its elements among factors of S-T. If D'_{r+1} is justified by being in T', again as above U_{r+1} consists of factors of elements of T, and, therefore, the last element of U_{r+1} has T-support.

If D'_{r+1} is a resolvent of D'_j and D'_k, at least one of D'_j and D'_k has T'-support since the ground deduction has T'-support. Say without loss of generality that D'_k has T'-support. But then by induction the last line B of U_k has T-support. Also by induction A, the last element of U_j, either has T-support or is a factor of a clause in S-T. U_{r+1} by construction consists either of C or C and D, where C is a resolvent of A and B and D (if needed) is a factor of C. In either case U_{r+1} has all of its elements with T-support.

If D'_{r+1} is justified as a paramodulant of two earlier clauses A and B, and if B has a term in a position corresponding to the term (on the ground level) of paramodulation, we can parallel the arguments just given for D'_{r+1} when D'_{r+1} is a resolvent. If D'_{r+1} is justified by paramodulation from D'_j into D'_k where D'_k has T'-support, and if B is the last element of U_k but does not have a term in the corresponding position to that of the term of paramodulation in D_k' then the procedure gives as U_{r+1} the sequence $G_1, G_2, \ldots, G_p, B_1, B_2, \ldots, B_p, C$ or possibly $\ldots C, D$. $G_1, G_2, \ldots, G_p$ are in S. By induction B has T-support since D_k' does. Therefore $B_1, B_2, \ldots, B_p$ have T-support. By induction the last element A of U_j either is a factor of an element of S-T or has T-support, so C has T-support. If the procedure calls for an application of Lemma 7 to yield a clause D, then D has T-support. The last line of U_{r+1}, therefore, has T-support.

If D_{r+1}' has T'-support and is a paramodulant of D_j' into D_k', but D_k' does not have T'-support, then D_j' has T'-support since the ground deduction is T'-supported. We can parallel the argument just given in the previous paragraph and conclude that U_{r+1} consists either of factors of clauses of S-T or clauses with T-support and that the last element of U_{r+1} has T-support. The induction is, therefore, complete, and we have the proof of Theorem 1.

Since S'-T' was shown to be R-satisfiable, the empty clause must have T'-support. D_h, therefore, must have T-support since D_h is the last element of U_n.

(It should not be concluded that a completeness theorem for paramodulation without functional reflexivity would lead directly to the completeness theorem for paramodulation with set of support in the absence of functional reflexivity. The functional reflexivity axioms were used directly in the proof of Theorem 1.)

<u>Theorem 2.</u> If S is an unsatisfiable set of (not necessarily ground) clauses, and if $T \subseteq S$ is such that S-T is satisfiable, then $S \vdash_{\Sigma T} \square$; set of support is refutation complete within ΣT.

Proof. The proof parallels that just given for Theorem 1, omitting all references to functional reflexivity, paramodulation, and replacing R-satisfiability by satisfiability, etc.

It is perhaps interesting to note that the proof of Theorem 1 easily yields refutation completeness within ΣT, i.e., completeness of set of support for resolution. The previously given proofs for refutation completeness in ΣT [9][10], however, do not seem to generalize easily to refutation completeness in ΠT.

Corollary 3. For finite functionally reflexive sets there is a refutation procedure for Π with set of support. That is, there is a uniform procedure that will, given any T and a finite functionally reflexive R-unsatisfiable $S \supseteq T$ with S-T R-satisfiable, generate (in a finite number of steps) a refutation of S in the system ΠT.

Proof. By Theorem 1 there exists a T-supported refutation $D_1, D_2, \ldots, D_k$. Since by definition refutations are finite in length and the D_i either are factors of S-T or are in some T_S^j, there exists an n such that $S^0 \cup T_S^n$ contains all D_i. Consider the procedure which generates $(S-T)^0$, then $T_S^0, T_S^1, \ldots, T_S^n$. Since S is finite, $(S-T)^0$ and T_S^j for $0 \leq j \leq n$ are all finite, the given procedure will find the T-supported refutation of S after generating only finitely many clauses.

References

1. Andrews, P. "Resolution with Merging." *J. ACM* 15 (1968), pp. 367-381.

2. Darlington, J. "Theorem-proving and Information Retrieval." *Machine Intelligence IV* (1969), ed. by D. Michie and B. Meltzer.

3. Green, C. "Theorem-proving by Resolution as a Basis for Question-answering Systems." *Machine Intelligence IV* (1969), ed. by D. Michie and B. Meltzer.

4. Robinson, G., Wos, L., and Carson, D. "Some Theorem-proving Strategies and Their Implementation," AMD Technical Memorandum #72, Argonne National Laboratory (1964).

5. Robinson, G., and Wos, L. "Completeness of Paramodulation." (Abstract), *J. Symb. Logic,* 34 (1969), p. 160.

6. Robinson, G., and Wos, L. "Paramodulation and Theorem-proving in First-order Theories with Equality." *Machine Intelligence IV* (1969), ed. by D. Michie and B. Meltzer, pp. 135-150.

7. Robinson, J. "A Machine-oriented Logic Based on the Resolution Principle." *J. ACM* 12 (1965), pp. 23-41.

8. Robinson, J. "Automatic Deduction with Hyper Resolution." *Internat. J. Assoc. Comput. Math.* 1 (1965), pp. 227-234.

9. Slagle, J. "Automatic Theorem Proving with Renamable and Semantic Resolution." *J. ACM* 14 (1967), pp. 687-697.

10. Wos, L., Robinson, G., and Carson, D. "Efficiency and Completeness of the Set of Support Strategy in Theorem Proving." *J. ACM* 12 (1965), pp. 536-541.

11. Wos, L., and Robinson, G. "The Maximal Model Theorem." (Abstract), *J. Symb. Logic,* 34 (1969), pp. 159-160.

12. Wos, L. and Robinson, G. "Maximal Models and Refutation Completeness." (Unpublished)

Offsetdruck: Julius Beltz, Weinheim/Bergstr.